Advancing Global Bioethics

Volume 5

Series editors

Henk A.M.J. ten Have
Pittsburgh, USA

Bert Gordijn
Dublin, Ireland

The book series Global Bioethics provides a forum for normative analysis of a vast range of important new issues in bioethics from a truly global perspective and with a cross-cultural approach. The issues covered by the series include among other things sponsorship of research and education, scientific misconduct and research integrity, exploitation of research participants in resource-poor settings, brain drain and migration of healthcare workers, organ trafficking and transplant tourism, indigenous medicine, biodiversity, commodification of human tissue, benefit sharing, bio-industry and food, malnutrition and hunger, human rights, and climate change.

More information about this series at http://www.springer.com/series/10420

Alireza Bagheri • Jonathan D. Moreno
Stefano Semplici
Editors

Global Bioethics: The Impact of the UNESCO International Bioethics Committee

Editors
Alireza Bagheri
Tehran University of Medical Sciences
Tehran, Iran

Jonathan D. Moreno
University of Pennsylvania
Philadelphia, USA

Stefano Semplici
Faculty of Humanities and Philosophy
University of Rome "Tor Vergata"
Rome, Italy

ISSN 2212-652X   ISSN 2212-6538   (electronic)
Advancing Global Bioethics
ISBN 978-3-319-22649-1   ISBN 978-3-319-22650-7   (eBook)
DOI 10.1007/978-3-319-22650-7

Library of Congress Control Number: 2015953502

Springer Cham Heidelberg New York Dordrecht London
© Springer International Publishing Switzerland 2016
This work is subject to copyright. All rights are reserved by the Publisher, whether the whole or part of the material is concerned, specifically the rights of translation, reprinting, reuse of illustrations, recitation, broadcasting, reproduction on microfilms or in any other physical way, and transmission or information storage and retrieval, electronic adaptation, computer software, or by similar or dissimilar methodology now known or hereafter developed.
The use of general descriptive names, registered names, trademarks, service marks, etc. in this publication does not imply, even in the absence of a specific statement, that such names are exempt from the relevant protective laws and regulations and therefore free for general use.
The publisher, the authors and the editors are safe to assume that the advice and information in this book are believed to be true and accurate at the date of publication. Neither the publisher nor the authors or the editors give a warranty, express or implied, with respect to the material contained herein or for any errors or omissions that may have been made.

Printed on acid-free paper

Springer International Publishing AG Switzerland is part of Springer Science+Business Media (www.springer.com)

In the last twenty years, I believe this is perhaps UNESCO's greatest achievement – our relentless determination to link ethics with science, to never let scientific development outstrip our ability to weigh it critically against the only standard that matters.

Irina Bokova
Director-General of UNESCO.

Preface

Since the beginning of the twenty-first century, discussions about 'global bioethics', its definition, methodology and application have generated a large body of literature in bioethics.

The UNESCO International Bioethics Committee (IBC), established in 1993, quickly became a major influence in advancing the global dialogue in bioethics and continues to advocate a more balanced and culturally less biased approach to bioethics. Since then the IBC has played a leading role in developing normative instruments as well as promoting a broader concept of bioethics. The members of the IBC have worked to bring global and local perspectives closer together and to present a shared understanding of the main values and principles of bioethics. UNESCO's international declarations, guidelines and reports have been instrumental in shaping the global bioethics discussion and have inspired member states to embrace the spirit of these instruments in national legislation, bioethics-related guidelines and public policies. Those normative instruments are seen as fulfilling the standard-setting mission of UNESCO.

Although there are various other international standards for bioethics-related practices, such as the World Medical Association's Declaration of Helsinki for human research ethics, the IBC has been the only regular, sustained forum for the interaction of persons from several dozen member states. The benefits of this interaction are not always tangible. Although they do not technically "represent" their countries, IBC members have established an informal network of collegiality and friendship that opens up a unique forum for the exchange of views and an enhanced appreciation for the challenges facing bioethics workers in different cultural milieus.

This book presents a review of the evolving global bioethical discussions and describes the reflections of the IBC on the most critical topics as well as the positions taken by the IBC in leading the global bioethical discussions. The contributors are mostly members or former members of the IBC, and the topics cover the conceptual premises of a universal framework for bioethics and the IBC's efforts in bioethical normative setting at the global level. Several chapters look at current IBC initiatives and discuss the impact of IBC initiatives on bioethics capacity building at national and regional levels in different parts of the world. Yet other chapters present new frontiers requiring thoughtful bioethical discussions.

In the first chapter, Henk ten Have elaborates on the achievements of two decades of the IBC's involvement in bioethical discussion at the global level. He argues that more important than contributing to the adoption of normative instruments, the IBC has played a leading role in promoting a broader conception of bioethics that is more appropriate to current processes of globalization. By explaining why an international organization such as UNESCO should be involved in global bioethics, the author presents some new bioethical challenges which the IBC should tackle in the next 20 years.

In Chap. 2, Michèle Stanton-Jean examines the importance of UNESCO's declarations on different bioethical issues in global bioethics. As an example, she elaborates the Universal Declaration on Bioethics and Human Rights by examining its elaboration, implementation, promotion and contribution to knowledge construction. The author concludes that the declarations provide value to global discussion and practice, especially in countries where bioethical infrastructures were previously absent.

By elaborating the key points in the success of the IBC's contributions over the past 20 years to bioethics at the global level, Nouzha Guessous emphasizes that the defining characteristic of all successful initiatives is the overarching multidisciplinary and pluralist approach of the IBC.

The article highlights the leadership of UNESCO and the IBC in global bioethics discussions and suggests a list of priorities for the UNESCO bioethics programmes.

Chapter 4, by Richard Magnus, discusses the universality of the Universal Declaration on Bioethics and Human Rights. The article explains how this declaration has shifted the main focus of bioethics from respect for individual autonomy to consideration of the good for the larger society. He advocates that we must think even more broadly to look at the consequences to humanity and beyond, including our future generations, the environment and other living beings. The author further submits that the declaration has broadened the scope and impact of bioethics by integrating international human rights law into the field of biomedicine.

In Chap. 5 based on the Universal Declaration on Bioethics and Human Rights Sheila McLean focuses on the importance of informed consent in global bioethics. She argues that while the declaration seeks to establish a normative framework, the IBC's further work contained in the report on consent helps guide states in appreciating how these norms might be translated into their culture and laws.

Next, in Chap. 6 Stefano Semplici emphasizes the "*social*" dimension of bioethics" and elaborates on the broader scope of bioethics. He argues that inasmuch as bioethics is about health and healthcare, it is at the very crossroads of all the determinants of human development and well-being. The author reflects on a sustainable holistic approach in which global bioethics should be understood as social bioethics and everyone should act accordingly, whether at the domestic or international levels.

In Chap. 7, Emilio La Rosa reflects on the report of the ICB on Traditional Medicine Systems and their Ethical Implications and presents ethical challenges in the application of traditional medicine. He argues that traditional medicine must not be an alternative for the poor, nor should it be a pretext for failing to improve access to the best diagnostic techniques and treatment. He further submits that traditional and modern medicine can coexist provided bridges are built between them. The author criticizes efforts to develop a two-tier healthcare system; rather, there should be one system that is easy to access and inexpensive for all.

Chapter 8 touches upon the complexity of ethical issues in biobanking. Ewa Bartnik and Eero Vuorio outline some ethical concerns in the systematic collection of human samples and data in biobanks. After providing a balanced account of the risks and benefits of biobanking, the authors offer practical approaches to population bioethics as well as how to deal with incidental findings.

Alireza Bagheri examines some of the ethical issues in organ transplantation and trafficking in Chap. 9. By elaborating the risks of stigmatization in organ procurement as well as the risks of discrimination in organ allocation, the author recalls the report of the IBC on the Principle of Non-Discrimination and Non-Stigmatization and outlines some practical measures to prevent stigmatization and discrimination in organ transplantation.

Abdallah Daar and his colleague examine the topic of nanotechnology, specifically the ethical, economic, environmental, legal and social issues concerning its development and application in Chap. 10. In reviewing the advances in nanotechnology that are most likely to benefit low- and middle-income countries, they examine the most relevant ethical challenges and warn about the "nano-divide" between high-income countries and the developing world. The authors propose potential approaches to address these challenges based upon the foundations of equity, justice, non-discrimination and non-stigmatization as advanced in the report of the IBC on the Principle of Non-Discrimination and Non-Stigmatization.

Jean Martin, in Chap. 11, examines Article 19 of the Universal Declaration on Bioethics and Human Rights, which calls for the establishment of ethics committees at various levels. He elaborates the conditions and rules necessary for national bioethics committees in order to comply with the requirements of independence, multidisciplinarity and pluralism.

In Chap. 12 Christiane Druml examines the influence of UNESCO's bioethics initiatives in Europe and elaborates on the interaction between the IBC and the network of ethical advisory bodies in Europe as an example. The author argues that such influence should be evaluated in a different way compared to other regions. She emphasizes the importance of the interaction and influence of the European members of the IBC with their various national ethical bodies as well as academia in Europe.

Olga Kubar and Jože Trontelj present a review of bioethics development in Central and Eastern Europe in Chap. 13. They make the case that the great historical and economical changes over the last 20 years – coinciding with IBC activities – have created unique opportunities for capacity building in bioethics. The authors argue that the formation of the Commonwealth of Independent States, uniting 11 regional countries, gave rise to a dynamic legislative and administrative collaboration in biomedical ethics with special efforts focusing on the implementation of IBC declarations.

The impact of the UNESCO bioethics programmes on the development of bioethics in Arabic countries is the topic of Chap. 14. Ben Ammar and his colleague examine how these programmes have shaped and impacted bioethics development in the Arab region. The authors argue that the core bioethical principles which have been emphasized by the UNESCO bioethics declaration are in harmony with Islamic values.

Chapter 15 authors Claude Vergès De Lopez and colleagues discuss the impact of the IBC activities on bioethics development in Latin America. They emphasize how the IBC's central focus on respect for cultural diversity, pluralism and human rights has been an important contribution to Latin American bioethics. The authors explain the positive impact of the Universal Declaration on Bioethics and Human Rights on bioethical issues and especially its influence on the interpretation of laws relating to health services in Latin America.

In recent years, Africa has become the focus of UNESCO's programmes. In Chap. 16 Monique Wasunna and her colleagues describe how IBC initiatives and documents in this area have been helpful in bioethics capacity building in Africa over the last two decades. The authors conclude that UNESCO bioethics initiatives and programmes have contributed immensely to the development of bioethics in Africa by supporting the establishment of national bioethics committees, strengthening the capacity of these committees, training teachers in bioethics and providing ongoing direction in addressing bioethics issues in the life sciences.

In Chap. 17 Myongsei Sohn elaborates on the development of bioethics in East Asia and the impact of the IBC's work on that region. He explains how the region, once an importer of Western bioethics, has developed its own bioethics discourse and has become a global contributor to the bioethics discussion.

In his address on the occasion of the inauguration and first session of the International Bioethics Committee, Federico Mayor, Director-General of UNESCO, pointed out the task to perform: "…the IBC is envisaged first of all as a forum for the exchange of ideas. It will also, as a corollary, be the inspiration for practical actions to be carried out in the field. Far from being set up as a monitoring agency to censure and stigmatize, its central task will be to facilitate understanding of the changes currently occurring, taking account of cultural differences. It will endeavour to heighten awareness and to inform the public and finally, will seek to foster dialogue between the representatives of specialist circles throughout the world, without losing sight of the fact that bioethics is something that directly concerns public decision-makers". Twenty years later, this commitment is still key for all those who want to contribute to foster the awareness as well as the instruments to respect, protect and fulfil the fundamental unity of humankind.

Finally, the editors would like to thank our IBC colleagues for their scholarly contributions to this book. They have successfully provided an in-depth analytical review of the IBC activities as well as its leading role in global bioethical discussion.

Our thanks are due to the anonymous reviewers whose comments were very useful to improve the quality of the discussion in each chapter. We also would like to acknowledge the help of the IBC Secretariat in translating two articles into English.

Tehran, Iran Alireza Bagheri
Philadelphia, USA Jonathan D. Moreno
Rome, Italy Stefano Semplici

Contents

1. **Globalizing Bioethics Through, Beyond and Despite Governments** 1
   Henk ten Have

2. **The UNESCO Universal Declarations: Paperwork or Added Value to the International Conversation on Bioethics? The Example of the Universal Declaration on Bioethics and Human Rights** 13
   Michèle Stanton-Jean

3. **Twenty Years of the International Bioethics Committee: Achievements and Future Priorities** 23
   Nouzha Guessous

4. **The Universality of the UNESCO *Universal Declaration on Bioethics and Human Rights*** 29
   Richard Magnus

5. **Consent and Bioethics** 43
   Sheila A.M. McLean

6. **Global Bioethics as *Social* Bioethics** 57
   Stefano Semplici

7. **Ethics and Traditional Medicine** 73
   Emilio La Rosa Rodríguez

8. **Biobanks: Balancing Benefits and Risks** 81
   Ewa Bartnik and Eero Vuorio

9. **The Risk of Discrimination and Stigmatization in Organ Transplantation and Trafficking** 91
   Alireza Bagheri

10	**Dust of Wonder, Dust of Doom: A Landscape of Nanotechnology, Nanoethics, and Sustainable Development** .... Fabio Salamanca-Buentello and Abdallah S. Daar	101
11	**The National Bioethics Committees and the Universal Declaration on Bioethics and Human Rights: Their Potential and Optimal Functioning** .... Jean F. Martin	125
12	**The UNESCO International Bioethics Committee and the Network of Ethical Advisory Bodies in Europe: An Interactive Relationship** .... Christiane Druml	137
13	**The Impact of the UNESCO International Bioethics Committee's Activities on Central and Eastern Europe** .... Olga Kubar and Jože Trontelj	145
14	**Bioethics in Arab Region and the Impact of the UNESCO International Bioethics Committee** .... Sadek Beloucif and Mohamed Salah Ben Ammar	151
15	**The Impact of the UNESCO International Bioethics Committee on Latin America: Respect for Cultural Diversity and Pluralism** .... Claude Vergès De Lopez, Delia Sánchez, Volnei Garrafa, and Andrés Peralta-Corneille	163
16	**Bioethics Development in Africa: The Contributions of the UNESCO International Bioethics Committee** .... Monique Wasunna, Aïssatou Toure, and Christine Wasunna	175
17	**Bioethics in East Asia: Development and Issues** .... Myongsei Sohn	185

Index 197

About the Editors

Alireza Bagheri-Chimeh is Assistant Professor of Medicine and Medical Ethics in Tehran University of Medical Sciences, Iran. He serves as the Vice-chairman of the UNESCO International Bioethics Committee and a member of the board of directors of the International Association of Bioethics. He is a senior editor of the Encyclopedia of Islamic Bioethics and a member of Editorial Board of the Encyclopedia of Global Bioethics.

Jonathan D. Moreno is the David and Lyn Silfen University Professor of Ethics, Professor of Medical Ethics and Health Policy, of the History and Sociology of Science, and of Philosophy at the University of Pennsylvania. He is a member of the UNESCO International Bioethics Committee and an elected member of the Institute of Medicine of the U.S. National Academy of Sciences.

Stefano Semplici is the current Chairperson (2011–2015) of the International Bioethics Committee of UNESCO. He is professor of Social Ethics at the University of Rome "Tor Vergata", editor of the international journal "Archivio di Filosofia-Archives of Philosophy" and Scientific Director of the "Lamaro Pozzani" College, Rome.

Contributors

Alireza Bagheri is Assistant Professor of Medicine and Medical Ethics in Tehran University of Medical Sciences, Iran. He serves as the Vice-chairman of the UNESCO International Bioethics Committee and a member of the board of directors of the International Association of Bioethics.

Sadek Beloucif is professor of Anesthesiology and Critical Care Medicine in Paris University. He has been a member of the French National Ethics Committee and president of the Advisory Council of the French Agence de la Biomédecine.

Mohamed Salah Ben Ammar is professor of Anesthesiology and Critical Care Medicine in El Manar University, Tunis. He is a member of the Tunisian National Ethics Committee and a member of the UNESCO International Bioethics Committee.

Ewa Bartnik is Professor of Molecular Biology and Human Genetics, University of Warsaw, and Professor, Institute of Biochemistry and Biophysics, Polish Academy of Sciences. She is Vice-chairperson of the UNESCO International Bioethics Committee.

Abdallah S. Daar is Professor of Public Health Sciences and Professor of Surgery, University of Toronto, Canada and a Member, Scientific Advisory Board of the Secretary-General of the United Nations. He is a fellow of the Third World Academy of Science (TWAS) and a member of the UNESCO International Bioethics Committee.

Christiane Druml is Vice-Rector of the Medical University of Vienna and Chairperson of the Austrian Bioethics Commission. She is Vice-President of the Medical Council of the Republic of Austria and a member of the UNESCO International Bioethics Committee.

Volnei Garrafa is director, UNESCO Chair in Bioethics and Coordinator of the Post-Graduate Program in Bioethics, University of Brasilia. He is editor-in-chief, Brazilian Journal of Bioethics and executive Secretary of UNESCO Latin American and Caribbean Bioethics Network. He is a member of the UNESCO International Bioethics Committee.

Nouzha Guessous is Professor of Hassan II University of Casablanca, Morocco. Researcher and consultant in Bioethics and Human Rights and Associate Researcher to Centre Jacques Berque. She is a former member and Chair of the UNESCO International Bioethics Committee.

Henk ten Have is Director of the Center for Healthcare Ethics at Duquesne University in Pittsburgh, USA. He was former Director of the Division of Ethics of Science and Technology in UNESCO, Paris (2003–2010). He is Editor-in-Chief of the journal of Medicine, Philosophy and Healthcare.

Olga Kubar is Head of the Clinical Department, Pasteur Institute, Saint-Petersburg. She is a Former Chairperson, Forum for Ethics Committees in the Commonwealth of Independent States. She is a member of the UNESCO International Bioethics Committee.

Emilio La Rosa Rodríguez is a Surgeon and Doctor in Anthropology and Human Ecology and a member of the Peruvian Society of Bioethics. He is former Director, Health and Society Study and Research Centre, France, and a former member of the UNESCO International Bioethics Committee.

Richard Magnus is Chief Judge (Ret) and Chairperson of the Bioethics Advisory Committee of Singapore. He is a member of the UNESCO International Bioethics Committee and Singapore's representative to the ASEAN Intergovernmental Commission on Human Rights.

Jean F. Martin is a public health physician and a founding member of the Swiss National Commission on Biomedical Ethics and a Member of Honor of the Swiss Public Health Association and a former member of the UNESCO International Bioethics Committee.

Sheila A.M. McLean is the first holder of the International Bar Association Chair of Law and Ethics in Medicine at Glasgow University. She has acted as a consultant/adviser to the World Health Organisation, the Council of Europe, UNESCO and a number of individual states and is a former member of the UNESCO International Bioethics Committee.

Jonathan D. Moreno is the David and Lyn Silfen University Professor of Ethics, Professor of Medical Ethics and Health Policy, of the History and Sociology of Science, and of Philosophy at the University of Pennsylvania. He is a member of the

UNESCO International Bioethics Committee and an elected member of the Institute of Medicine of the U.S. National Academy of Sciences.

Andrés Peralta-Corneille is Professor of Bioethics in Santiago Technological University, Dominican Republic and Founding member and former Chairperson of the National Bioethics Committee. He is a former member if the UNESCO International Bioethics Committee.

Fabio Salamanca-Buentello is a physician from the Faculty of Medicine of the National University of Mexico with a MSc in Bioethics from the Joint Centre for Bioethics. He is a PhD candidate in the ethical, social, and cultural components of global mental health in University of Toronto, Canada.

Delia Sánchez is former Assistant Professor of Bioethics, School of Medicine, University of Uruguay and former director of Health Technology Assessment Unit, Ministry of Health She is Academic Coordinator of MERCOSUR Health Systems Observatory and is Vice-chairperson of the UNESCO International Bioethics Committee.

Stefano Semplici is the current Chairperson (2011–2015) of the International Bioethics Committee of UNESCO. He is professor of Social Ethics at the University of Rome "Tor Vergata", editor of the international journal "Archivio di Filosofia-Archives of Philosophy" and Scientific Director of the "Lamaro Pozzani" College, Rome.

Myongsei Sohn is Director of Asian Institute for Bioethics and Medical Law and Founding Chairperson of the Department of Bioethics and Health Law of Yonsei University, South Korea. He is a member of the UNESCO International Bioethics Committee

Michèle Stanton-Jean is Visiting Scholar, Centre de recherche en droit public, Faculté de Droit, Universite de Montréal. She was former Chairperson of the UNESCO International Bioethics Committee (2002–2005).

Jože Torntelj is Professor of Neurology and President of the Slovenian Academy of Sciences and Arts. He is Chairperson of the National Medical Ethics Committee and a member of the Steering Committee on Bioethics of the Council of Europe. He is a former member of the UNESCO International Bioethics Committee.

Aïssatou Toure is Immunologist and researcher in Pasteur Institute, Dakar and a member of the National Health Research Council, Senegal. She is a former member of the UNESCO International Bioethics Committee.

Claude Vergès De Lopez is Professor of Deontology and Bioethics, Faculty of Medicine, University of Panama and a member, Committees on Research Ethics,

Hospital del Niño and *Instituto Gorgas de Estudios en Salud*. She is a member of the UNESCO International Bioethics Committee

Eero Vuorio is Professor of Molecular Biology, University of Turku and Director of Biocenter Finland. He is former Chairperson of the National Board of Research Ethics and a member of the UNESCO International Bioethics Committee

Monique Wasunna is Consultant Physician and Specialist in Tropical Medicine and Infectious Disease and Assistant Director (Research) of the Kenya Medical Research Institute (KEMRI). She is a member of the Kenya National Bioethics Committee and Vice-Chairperson of UNESCO International Bioethics Committee.

Christine Wasunna is Principal Research Scientist and a member of the Biotechnology Research Programme at the Centre for Clinical Research, Kenya Medical Research Institute (KEMRI). She holds a PhD degree in Molecular Biology and a Postgraduate Diploma in Health Research Ethics.

Chapter 1
Globalizing Bioethics Through, Beyond and Despite Governments

Henk ten Have

Abstract This chapter will review the role of UNESCO and in particular the International Bioethics Committee (IBC) in the field of bioethics over the past two decades. Three questions will be addressed. The first question is what has been achieved. It will be argued that more important than contributing to the adoption of normative instruments the IBC has played a leading role in promoting a broader conception of bioethics that is more appropriate to current processes of globalization. Another question is why an organization such as UNESCO is involved, and should be more involved in the global development of bioethics. The last question that will be discussed concerns the challenges that will lie ahead in the next 20 years.

1.1 Introduction: What Has Been Achieved?

It is easy to enumerate the list of unique achievements of the bioethics program of UNESCO. The adoption of three normative instruments, the establishment of the IBC (with independent scientific experts as members) and the Intergovernmental Bioethics Committee (with governmental representatives as members), the creation of national bioethics committees in several countries, the promulgation of the core bioethics curriculum in universities around the world, and the setting up of the Global Ethics Observatory with data on bioethics experts, ethics entities, ethics teaching programs, and ethically relevant legislation in various Member States are all major achievements that help to promote and sustain bioethics across the world.

Without the activities of UNESCO these achievements would not exist today. However, these results and activities are the manifestations of a deeper concern that is closely related to the mission of UNESCO: the intellectual and moral solidarity of humanity that is the only guarantee that progress in science will contribute to human flourishing, peace and security. Against this foundational background that

---

H. ten Have (✉)
Center for Healthcare Ethics, Duquesne University, Pittsburgh, PA, USA
e-mail: tenhaveh@duq.edu

will be elaborated in the next paragraph, UNESCO's early involvement in bioethics can be explained. By 1970, the organization had started to organize symposia and conferences on bioethics, mainly related to the development of genetics, life sciences and reproductive technologies and in cooperation with UNESCO's Scientific Coordinating Committee for the Human Genome Project (UNESCO 1993). It is also remarkable that the Council for International Organizations of Medical Sciences (CIOMS) established in 1949 jointly by WHO and UNESCO with the purpose of promoting international activities in the biomedical sciences convened in 1967 a conference in Paris (most probably in UNESCO's Headquarters), on biomedical science and the dilemma of human experimentation. This conference would be the first of a long series of annual Round Tables on the ethical aspects of modern medicine, organized with the assistance of UNESCO and WHO (Bankowski and Dunne 1982). In 1976 CIOMS appointed a standing Advisory Committee on Bioethics. This Committee started an intensive international dialogue among researchers, ethics experts and policy-makers from around the world in order to develop guidelines for international medical research.

As parent organization and co-organizer UNESCO must have become aware of the increasing ethical issues associated with the rapid development of medicine and life sciences, even before the term "bioethics" was coined. While CIOMS membership consists of academic and scientific organizations, UNESCO is an intergovernmental organization; it brings together governments that have an interest in the promotion of science. In 1970, the word 'bioethics' was introduced for the first time by Van Rensselaer Potter who gave a broad definition of the concept (Ten Have 2012). The same era witnessed revolutionary changes and innovations in medical diagnosis and treatment but also in science and technology. Additionally, scandals, misuses and injustices came to light and alarmed the public and policy-makers, leading to the establishment of the first bioethics centers, ethics committees, review boards and efforts to codify patient rights and to regulate the medical and biomedical research community (Rothman 1991). Although bioethics emerged first in the United States, it quickly developed in other parts of the world. However, the birth and early growth of bioethics left conceptual and methodological markers on the new discipline. The story is familiar. Potter complained that the new term 'bioethics' was readily accepted and widely used but only as a new fancy name for 'medical ethics,' not as a different approach to ethical issues. The acceptance of the term falsely suggested that there was something new, as he himself intended by coining the term, but in fact it only continued the traditional approach albeit under a different guise (Potter 1988). The approach to bioethics that became dominant, especially under the influence of the Kennedy Institute at Georgetown University (established in 1971, with 'bioethics' in its original name) has two flaws according to Potter. First, it is concerned with the perspective of the individual patient. Its main concern is how individual lives could be enhanced, maintained, and prolonged through the application of medical technologies. Second, it is exclusively interested in the short-term consequences of medical and technological interventions. It is not concerned with what Potter regarded as the basic and most urgent ethical problems of humankind that are threatening the human survival, problems such as poverty, pollution, and violence. These flaws assure that contemporary bioethics is not generating really new perspectives and new syntheses that are focused on safeguarding the future of the human species. In order to reiterate that a new approach is essential, Potter again introduced in the late 1980s a new term: *global bioethics* (Potter 1988). What we currently have is biomedical ethics or medical bioethics. What we need is an unquestionably innovative and multi-disciplinary approach that combines medical and individual perspectives with social and ecological concerns so that it has global scope, not only in the sense that it has worldwide significance but also is encompassing and broad (Ten Have and Gordijn 2013).

The peculiarities and characteristics of mainstream bioethics have become under particular scrutiny during the 1990s. It was recognized that the growing appeal of this new discipline among public and scientific circles of opinion leaders could be attributed to the empowering combination of two traditional notions from the history of moral philosophy: 'application' and 'principle'. In Beauchamp and Childress' well-known textbook, biomedical ethics is defined as *applied ethics*, "the application of general ethical theories, principles and rules to problems of therapeutic practice, health care delivery, and medical and biological research" (Beauchamp and Childress 1983: ix–x). Instead of the theoretical abstractions of traditional moral philosophy, applied ethics can contribute to the analysis of dilemmas, the resolution of complex cases and the clarification of practical problems arising in the healthcare setting. The practical usefulness of applied ethics not only manifests itself in biomedicine, but it has a wider scope; the same approach is important for other areas such as business ethics and environmental ethics. Applied ethics therefore can extend to almost any area of life where ethical issues arise. 'Application' here has a double connotation: it indicates that ethics is available for what we usually do, it applies to our daily problems; but it also is helpful, practical, in the sense that ethics is something to do; it works to resolve our problems.

The second characteristic of the dominant conception of bioethics is the focus on *principles*. If ethics is conceived as applied ethics, then subsequent reflection is needed on what is being applied. The emerging consensus that principles should provide the answer to this quest is coherent with the moralities of obligation that have dominated modern ethical discourse, especially since Kant. Behavior in accord with moral obligations is considered morally right. Morality is understood as a system of precepts or rules people are obliged to follow. Particularly in the early days of bioethics, when medical power was strongly criticized, and the rights of patients were vehemently emphasized as requiring respect, the moralities of obligation presented themselves as a common set of normative principles and rules that we are obliged to follow in practice. As Gracia (1999) has pointed out, the *Belmont Report* in 1978 was influential because it was the first official account to identify three basic ethical principles: autonomy, beneficence and justice. A basic principle was defined as a general judgment serving as a justification for particular prescriptions and evaluations of human actions. From these principles, ethical guidelines can be derived that could be applied to the biomedical area. About the same time, Beauchamp and Childress, in the first edition of their book, introduced the four-principles approach, adding "nonmaleficence" to the above three principles. In their view, principles are normative generalizations that guide actions.

Although Beauchamp and Childress have considerably elaborated and adapted their theoretical framework in later editions, their work has contributed to the conception of bioethics that has long dominated the practical context, in ethics committees, clinical case-discussions, ethics courses, and compendia and syllabi. This conception is sometimes called 'principlism' The focus is on the use of moral principles to address ethical issues and to resolve conflicts at the bedside (DuBose et al. 1994). Long-time editor of the *Journal of Medical Ethics*, Raanan Gillon, for example, argues that principlism is a universal tool; it provides a method to resolve all moral issues in all areas of daily life, whatever the personal philosophies, politics, religions, cultural traditions and moral theories of the persons involved (Gillon 1994). Although principlism was dominant approach in bioethics, it was not the only one and it was also criticized from the start. Since the 1990s this criticism has grown (Ten Have 2011). One critique is that it pays insufficient attention to the practical setting since it uses the mould of the four moral principles to address actual cases and issues, without taking into account the concretely lived experiences of patients and health professionals. The other critique points out that bioethics has developed within a particular Western cultural and social context while at the same time critical reflection upon the social and cultural value system within and through which it operates is rare. In response to these criticisms new approaches to bioethics became more influential during the 1990s, for instance phenomenological ethics, hermeneutic ethics, narrative ethics, and care ethics. It was also recognized that 'application' was often regarded in a restricted sense since bioethical debate was usually focused on a highly selective and limited number of topics and issues that were associated with individual choice and technological opportunities; these bioethical issues were relevant for Western countries but not for the majority of people living in the developing world. Finally, it became clear that the focus of principlism on duties and rights was associated with the dominance of the moral principle of individual autonomy while in other cultures more emphasis is placed on responsibility and community.

The development of mainstream bioethics as well as the critical discourses it engendered over the past few decades provides the background for new phenomena that have transformed bioethics recently. Processes of globalization have not only affected medicine, healthcare and research but at the same time bioethics. Bioethical discourse no longer is crossing borders and thus transnational; rather, it has become supra-territorial, i.e. relevant to all countries and taking into account the concerns of all human beings wherever they are. This means that for bioethics it is no longer sufficient to 'export' principlism to non-Western countries, for example, making translations of well-known textbooks in many languages, or providing fellowships for training and intensive courses in the U.S. or Europe. Rather, rethinking the approaches and methods of bioethics is required in order to address the new challenges of globalization. While bioethics may have primarily originated in Western countries, its reach and relevance are now planetary. On the one hand, the traditional issues of bioethics are confronted with new challenges. With the introduction of clinical trials in developing countries the concept of informed consent is confronted with different cultural traditions in which individual decision-making is an unusual concept. On the other hand, the existence of global markets has created new problems such as organ trade, medical tourism, corruption and bioterrorism. Even if such problems exist only in few countries, the way they are addressed will have consequences for other countries. Often national legislation or regulation will not be sufficient; instead, international cooperation and action will be required. Problems like pandemics, malnutrition, hunger and climate change require coordinated global policies and actions. Even if the moral values in specific countries and regions differ, a common ground has to be found as a world community. It is no longer sufficient to apply the restricted set of principles developed in the West. The same conclusion follows from Potter's use of the term 'global', referring to bioethics as more encompassing and comprehensive, combining traditional professional (medical and nursing) ethics with ecological concerns and the larger problems of society. This implies more than simply declaring that today's problems are global and affect everyone. First, it requires interdisciplinary cooperation. Global problems such as poverty, climate change and inequities in healthcare can only be addressed by obtaining and applying different types of knowledge. It is necessary to bridge the gap between science and humanities. Secondly, it requires that diverse perspectives must be used to explain and understand complex phenomena. Global problems can no longer be approached only from an exclusively Western (or Eastern or Southern) perspective; rather, they require a really global perspective.

The challenge to develop a global discourse of bioethics, and thus articulate an ethical framework that is focused on current processes of globalization in the area of medicine and healthcare, is particularly and uniquely faced by UNESCO. It is here that the major contribution of the organization and the IBC should be located.

1.2 Why UNESCO?

UNESCO is not a university but an intergovernmental organization with all the concomitant drawbacks of a complicated sometimes Kafkaesque bureaucracy, political dealing and wheeling, unexpectedly changing policies and vague, opaque compromises. It is clear that at least some in academia either don't understand the operations of such an organization because it seems too messy and politically tainted, or do not have much appreciation for its activities and achievements. Such a point of view is understandable from the perspective that bioethics is merely an academic enterprise of philosophers. However, the globalization of bioethics has thoroughly discredited this perspective. Global bioethics today is relevant for citizens everywhere; it has major impacts on the provision of healthcare, the management of pandemics, the development of new drugs and vaccines, and the improvement of living conditions. This does not mean that there is no need for academic teaching and research, or for philosophical reflection, but as multi-disciplinary discourse, global bioethics necessarily involves policy-makers, legislators, journalists, public opinion and business leaders as well as scientists and health professionals. There is a need to combine theoretical reflection with practical arrangements that are adapted to local circumstances and different settings. There is also a need to search for a middle ground between ethical foundationalism and anti-foundationalism, and thus associate universal discourse with respect for diversity (Ten Have 2011). Potter has emphasized the idea that bioethics essentially is a bridge, – not a final product but an activity that connects and brings together rather different points of view and various types of knowledge and experience so that it can benefit humanity as a whole (Potter 1971). UNESCO has assumed this role of bridge builder.

First, given its mission and its mandated work in the areas of education, culture, science and communication, the organization from the start emphasized its role as 'moral conscience.' This was clearly expressed in the vision of Julian Huxley, the first Director-General, that, in order to make science contribute to peace, security and human welfare, it was necessary to relate the applications of science to a scale of values. Guiding the development of science for the benefit of humanity therefore implied "the quest for a restatement of morality … in harmony with modern knowledge" (Huxley 1946). It is also reflected in the decision in June 1992 of Federico Mayor, then Director-General, to set up an International Bioethics Committee. Since member states have been particularly concerned about the relationship between scientific and technological progress, ethics and human rights, the Committee was asked to explore how an international instrument for the protection of the human genome could be drafted. The Committee met for the first time in September 1993. In the meantime, a Scientific and Technical Orientation Group was formed in December 1992, carrying out preparatory studies. The Group conducted extensive consultations, focusing on five themes: genome research, embryology, neurosciences, gene therapy, and genetic testing. For each theme various dimensions were studied: the current state of progress in research at the world level, the application of the results of this research, and the principal ethical concerns for the present and for the future. On the basis of these studies, the Group identified the reference points likely to secure the broadest agreement, proposing principles most likely to respond to the ethical concerns. The International Bioethics Committee then started its work and developed a proposal for a framework of ethical principles for a possible international instrument for the protection of the human genome. This approach illustrates that in global bioethics scientific expertise and ethical assessment need to be combined.

Secondly, given the nature of UNESCO as an inter-governmental organization, bioethics is not merely explained or clarified but it needs to be applied in policies in different member states. Bioethical principles are therefore 'declared' as relevant so that they can be applied by policy-makers in their domestic constituencies. Declarations not only assert a framework of ethical principles that reflect the commitments and value systems of all member states but they also provide indications about what is the 'ethical minimum' that should be considered in the respective countries that have adopted them. Even if they are not binding normative instruments like conventions, they nevertheless contribute to the body of international human rights law guiding citizens across the world. There is no other global entity that combines standard-setting and practical implementation activities in the field of bioethics. All its imperfections notwithstanding, UNESCO is unique as a global forum in connecting ethics, science and policy-making. This last feature is exemplified in the International Bioethics Committee. The institutional functioning of this committee within the organization, and at the same time its independent role as expert advisory body to the Director-General, highlights a unique aspect of global bioethics. It also demonstrates that bioethics is more than an academic enterprise; it needs to be translated and transferred into practical activities on the ground. The expertise assembled in the IBC reflects this liaison of brains and hands.

Third, UNESCO is a global platform for bioethics that brings together moral perspectives from all countries on an equal basis (at least in principle since in practice some member states will be more involved than others). This is easier said than done. In practice, it means extensive deliberation and negotiation. Those who state that bioethics declarations are the result of western moral imperialism have no idea how international organizations such as UNESCO are working in practice. Influential member states such as Brazil, China, India, Saudi Arabia and South Africa are very sensitive to this possible accusation. Any suggestion and proposal, for example from a European contributor will a priori be critically scrutinized and assessed for its global relevance. At the same time, the request to develop normative instruments frequently originates from developing countries. Since they have no, or weak, infrastructures in the field of bioethics, they often are concerned that they may not be able to participate in scientific and technological developments or that they will be confronted with the negative fall-outs. The developed countries all have sufficient infrastructure with ethics committees, relevant legislation and public debate. This difference in interest and relevancy is reflected in the drafting and negotiation of declarations where the latter group of countries is often working with existing, often domestic frameworks and the first group with the urge to expand these into a framework that is really inclusive and expansive so that it is relevant in their cultural, religious and historical setting. The result of this tension is that the major challenge is always to reconcile an aspiring universal discourse with respecting local diversity. This is continuously reflected in the work of the IBC. When the Committee started the process of drafting the Universal Declaration on the Human Genome and Human Rights it considered that cultural diversity should be taken into account while at the same time asserting universality (UNESCO 1993).

These bridging functions have enabled UNESCO to facilitate and promote a new type of bioethics, viz. global bioethics. In a field with many different national and international players and stakeholders, the organization plays a special role, providing a global platform that connects moral perspectives from a multitude of countries, combining standard-setting with practical implementation activities as well as scientific expertise and policy-making. But there is also an almost ideological role. More than other international organizations, UNESCO is driven by certain values that are utterly relevant for global bioethics. Specifically relevant is its focus on the common heritage of humankind. This focus directs attention to the common good. It expresses the basic idea that humanity needs more than exchanging commodities in a free market, and that humankind can only survive if it cares for global interests.

This care can only be accomplished on the basis of pragmatic solidarity and respect for diversity. The problem is that the organization does not take this ideological role as seriously as it should.

1.3 Challenges in the Next Two Decades

Future activities should be based on an analysis of the fundamental challenges. The major problems of global bioethics nowadays are related to structural injustices and social inequalities in health and health care. Current international clinical research insufficiently contributes to the alleviation of the global health burden and does not help to eliminate structural injustice. (Ganguli Mitra 2013). Medical anthropologist Paul Farmer concluded that "…the fundamental problem of our era [is]: the persistence of readily treatable maladies and the growth of both science and economic inequality" (Farmer 2005). The goal of bioethical activities in the global era should therefore be to address global health inequities and to reinsert a social commitment in healthcare, not as a business but as a human engagement.

These challenges are inherently associated with the emergence of global bioethics. Born in the 1970s, bioethics is traditionally conceived as a response to the power of medical science and technology. Patients felt overwhelmed with the possibilities of modern medicine; paternalistic professionals and the availability of technological interventions often seemed to dictate what would be done. Bioethics expanded as a public discourse empowering individual citizens and encouraging legislation in some areas as research, transplantation, reproduction and end-of-life care. It has rapidly evolved into a strong discipline with a clear conceptual and methodological framework and with the appropriate hardware of a scientific discipline: textbooks, journals, conferences, associations and educational programs. But this type of bioethics is closely connected to more developed countries that are confronted with scientific advances and technological innovations. It has therefore a specific agenda and scope that is often irrelevant for the majority of the world population living in less developed countries with limited or no access to healthcare, and with no benefits from the progress in science and technology. The globalization of healthcare and medical research has created a different context for bioethics. The major bioethical issues of today no longer have to do with the power of science and technology but with the power of money. Neoliberal market ideology has created increasing inequalities in health and healthcare. Because welfare safety nets and healthcare systems have been privatized and social protective mechanisms deregulated and minimized, healthcare has become even more inaccessible, and individuals, groups and populations are now more vulnerable than before. The United Nations Development Program concluded in 1999, "People everywhere are more vulnerable". Global bioethics has therefore emerged as a new type of discourse that specifically addresses the impact of globalization on citizens across the world. The traditional focus on advanced technologies, scientific research, and sophisticated healthcare is no longer sufficient; bioethics need to be expanded; it needs to take account the effects of globalization, focusing on the forgotten, the invisible and the ignored billions of people who are powerless and voiceless, and lack basic healthcare. The major contribution of UNESCO over the past two decades is that it has greatly contributed to this change in bioethical perspective. There now is an ethical framework that promotes, in the spirit of Potter, a broader view of bioethics, linking individual, social and environmental concerns. There is an agenda with issues and topics such as social responsibility, benefit sharing and protection of future generations. The challenge is to put this framework into practice. This will require practical but also theoretical efforts.

It is useful to bear in mind an expression that is used in developing countries: laws are made of paper, bayonets are made of steel. In other words, talking is good but acting is better. The bioethics program of UNESCO should therefore continue its efforts to implement the normative framework of the declarations with practical activities. However, they may be more targeted than in the past recognising that not all conditions are equally fruitful for the development of bioethics. Highly selective targeting of specific countries may be necessary, also because of budget limitations. But within the selected targets, a broad range of interconnected activities should be employed: fostering functional bioethics committees, stimulating active teaching programs, encouraging public debate, but at the same time monitoring and reporting on progress, so that bioethical country models can be publicized. The program should also more vigorously participate in global policy activities showing concern for including bioethics in debates about global health. It should be present in consultative and deliberative processes such as the revision of the Declaration of Helsinki, the revision of the CIOMS Guidelines, and the initiatives to draft a Framework Convention on Global Health for the post-2015 development agenda.

But these practical activities can only be effective in the long run if they are guided by intellectual analyses. If the development of global bioethics is intrinsically related to the processes of globalization, it needs to do more than simply facilitate and explain these processes; it should critically scrutinize them. Today's bioethical problems such as poverty, corruption, inequality, organ trade, medical tourism, care drain and bioterrorism are affecting the whole of humankind. They are produced by neo-liberal market policies that have exposed more people worldwide to more threats and hazards, and have decreased their capacities to cope. They often jeopardize the well-being of human beings by damaging and unjust structures and policies. However, bioethical discourse commonly uses the same basic assumptions as neoliberal globalism, arguing that vulnerability and inequality should be addressed and reduced through protecting and empowering individual decision-makers. It is understandable that bioethics is concerned with the fall-out of globalizing processes for individual persons. But using an individual focus abstracted from the social and political dimension of human existence, and neglecting the impact of market mechanisms on social life, will not allow bioethical policies and guidelines to redress the production of vulnerability and inequality. What is a symptom of the negative impact of a one-dimensional view of human beings is remedied with policies based on the same type of view. As long as the problematic conditions creating and reinforcing human vulnerability and inequality are not properly analyzed and criticized, bioethics will only provide limited palliation.

1.4 Conclusion

UNESCO and the IBC have a unique opportunity to further develop a new mindset that will be reflected in new approaches in global bioethics. Rather than managing the problems they should engage in critical analysis of the underlying mechanisms and introduce a new set of values such as solidarity, cooperation, sharing of benefits and global justice. Focusing on sophisticated technologies or complex issues will no longer be sufficient. What is imperative is the development of a social bioethics focusing on countering structural injustice, marginalization and exploitation of vulnerable populations. The next generation of bioethical problems has less to do with 'converging technologies' but rather with 'diverging benefits'. Taking global justice as its central focus will imply a critical approach towards the neoliberal model of globalization that is disseminated by other international organizations such as the World Bank and the International Monetary Fund. This will be difficult for an intergovernmental organization since it will imply critical analyses of policies and proposals promulgated by sister organizations in the same international system. It can only be done by engaging intellectuals from different parts of the world. It will imply stronger cooperation with selected and engaged NGOs. It demands providing opportunities for giving voice to 'bioethics from below.' Obviously, moral courage will be necessary to demonstrate that the future of humankind is not dependent on governments but is in fact the concern of all citizens of the world. But this is exactly why the Organization was established in the first place.

References

Bankowski, Z., and J.F. Dunne. 1982. History of the WHO/CIOMS project for the development of guidelines for the establishment of ethical review procedures for research involving human subjects. In *Human experimentation and medical ethics*, Proceedings of the XVth CIOMS Round Table Conference Manila, 13-16 September 1981, ed. Z. Bankowski and N. Howard-Jones, 441–452. Geneva: CIOMS.

Beauchamp, T.L., and J.F. Childress. 1983. *Principles of biomedical ethics*, 2nd ed. New York: Oxford University Press.

DuBose, E.R., R. Hamel, and L.J. O'Connell (eds.). 1994. *A matter of principles? Ferment in U.S. bioethics*. Valley Forge: Trinity Press International.

Farmer, P. 2005. *Pathologies of power. Health, human rights, and the new war on the poor*. Berkeley: University of California Press.

Ganguli Mitra, A. 2013. A social connection model for international clinical research. *American Journal of Bioethics* 13(3): W1–W2.

Gillon, R. (ed.). 1994. *Principles of health care ethics*. Chichester: John Wiley & Sons.

Gracia, D. 1999. History of medical ethics. In *Bioethics in a European perspective*, ed. H.A.M.J. ten Have and B. Gordijn, 17–50. Dordrecht/Boston/London: Kluwer Academic Publishers.

Huxley, J. 1946. *UNESCO. Its purpose and its philosophy*. Preparatory Commission of the United Nations Educational, Scientific and Cultural Organization. Paris: UNESCO.

Potter, V.R. 1971. *Bioethics: Bridge to the future*. Englewood Cliffs: Prentice-Hall.

Potter, V.R. 1988. *Global bioethics: Building on the Leopold legacy*. East Lansing: Michigan State University Press.

Rothman, D.J. 1991. *Strangers at the bedside. A history of how law and bioethics transformed medical decision making*. New York: Basic Books.

Ten Have, H. 2011. Foundationalism and principles. In *The SAGE handbook of health care ethics: Core and emerging issues*, SAGE handbook series, ed. Ruth Chadwick, Henk ten Have, and Eric Meslin, 20–30. London: SAGE Publications.

Ten Have, H. 2012. Potter's notion of bioethics. *Kennedy Institute of Ethics Journal* 22(1): 59–82.

Ten Have, H., and B. Gordijn. 2013. Global bioethics. In *Handbook of global bioethics*, ed. A.M.J. Henk, H. ten Have, and B. Gordijn, 3–18. Dordrecht: Springer Publisher.

UNESCO. 1993. *Study submitted by the Director-General concerning the possibility of drawing up an international instrument for the protection of the human genome*. General Conference, 27th Session (27 C/45). http://unesdoc.unesco.org/images/0009/000954/095428eo.pdf. Accessed 28 Dec 2013.

Chapter 2
The UNESCO Universal Declarations: Paperwork or Added Value to the International Conversation on Bioethics? The Example of the Universal Declaration on Bioethics and Human Rights

Michèle Stanton-Jean

Abstract In October 2005, the General Conference of UNESCO adopted the *Universal Declaration on Bioethics and Human Rights*. Since its adoption, the Declaration has been the object of many publications both positive as well as negative. This article contends that the Declaration, although not perfect, is a valuable addition to the Bioethical conversation. First it discusses the theoretical issues of universality, globalization and human rights. It then takes a pragmatic approach by considering its development, implementation, promotion and contribution to knowledge construction thereby demonstrating its usefulness, especially in countries where bioethical infrastructures were previously absent.

2.1 Introduction

The UNESCO Recommendations and Declarations propose to Member States principles or norms that are susceptible of inspiring national legislations, guidelines, or regulations and provide a common understanding of bioethical issues. Those normative instruments are seen as fulfilling the standard-setting mission of UNESCO.

In October 2005, the Commission on Social and Human Sciences at UNESCO discussed the text of the *Universal Declaration on Bioethics and Human Rights* (UDBHR). After a short presentation, the chair invited the participants from the Member States who wished to comment. Those who took the floor made some general comments about the text, raising some of the points they would have worded differently but they all concluded their statement by saying that they were all ready

M. Stanton-Jean (✉)
Centre de recherche en droit public, Faculté de Droit, Universite de Montréal, Montréal, QC, Canada
e-mail: michele.stanton.jean@sympatico.ca

© Springer International Publishing Switzerland 2016
A. Bagheri et al. (eds.), *Global Bioethics: The Impact of the UNESCO International Bioethics Committee*, Advancing Global Bioethics 5,
DOI 10.1007/978-3-319-22650-7_2

to recommend to the plenary the adoption of the Declaration. As the Chair of the International Bioethics Committee during the development of the project, I travelled to many countries to explain and discuss the text; sat in many meetings to listen to all the stakeholders involved in the consultation process; and reflected on all the work that had been done by the IBC members, the government experts and the secretariat. I thought that it was a great moment for Bioethics. For the first time a global political statement in the field of Bioethics was adopted by all member states of UNESCO.

2.2 Bioethics and UNESCO

The Constitution of the United Nations Educational, Scientific and Cultural Organization (UNESCO) came into force on 4 November 1946. The preamble gives a good indication of its mandate, "That since wars begin in the minds of men, it is in the minds of men that the defences of peace must be constructed; That ignorance of each other's ways and lives has been a common cause, throughout the history of mankind, of that suspicion and mistrust between the peoples of the world through which their differences have all too often broken into war" (UNESCO 1945).

Article I, Paragraph 2 states that the Organization will "recommend such international agreements as may be necessary to promote the free flow of ideas by word and image" (UNESCO 1945). Article IV, Paragraph B.4 mentions two categories of instruments that can be developed by the organization: "conventions and recommendations". However, it states that these instruments must be approved by the General Conference and submitted to Member States for their approval. A third category, "declarations", also exists. It should be noted that; this category was not explicitly mentioned in the Constitution, but has become quite common, especially in recent years (UNESCO 2007a).

Declarations are like recommendations but are named as such because of their importance. Declarations are adopted during the General Conference and because of that solemnity engage governments to implement them in their countries. Although they are part of the body of soft law, they contribute to the development of positive law and provide the scientific community and the general public with a tool to push their respective governments to act.

The first declaration was the *Declaration of Principles of International Cultural Cooperation*, adopted in 1966 on the occasion of the Organization's twentieth anniversary. The procedures to follow when drafting a Declaration were adopted by the General Conference at its 33th session (UNESCO 2012). The recommendations and the declarations propose to Member States principles or norms that are susceptible of inspiring national legislations, guidelines, or regulations and provide a common understanding of certain issues. Those normative instruments are seen as fulfilling the standard-setting mission of UNESCO. During the development of the UDBHR questions have been raised about the involvement of UNESCO in the field of bioethics. It is important to mention what the former Director General of UNESCO

Koïchiro Matsuura wrote: "From the beginning of the Organization's activities in this field [bioethics], UNESCO's General Conference decided to adopt a gradual and prudent approach based on the knowledge available on this complex subject matter, which lies at the interface of many disciplines. Furthermore it decided to take into account the diverse contexts (scientific, cultural, social and economic), in which ethical thinking unfolds in different parts of the world. This approach has led to two important legal consequences. The first is the use of the "declaration" rather than the convention or recommendation for the setting of standards in the field of bioethics. Three important declarations have so far been adopted by UNESCO in this respect. The second consequence is the articulation of broad principles and norms, which could be accepted by all Member States of UNESCO in view of the universal nature of the issues involved" (UNESCO 2007b). The recent developments in life sciences and especially in genetics have highlighted many ethical implications. Having been involved in ethics through its science mission, the Director General of UNESCO at that time, Federico Mayor, proposed that the Member States endorse a recommendation to prepare an international instrument to protect the human genome. Member States agreed and the International Bioethics Committee (IBC) was created in 1993. The Committee produced the first UNESCO declaration in Bioethics, the *Universal Declaration on the Human Genome and Human Rights* (1997). During the following years, two other declarations were produced: the *International Declaration on Human Genetic Data* (1993) and the *Universal Declaration on Bioethics and Human Rights* (2005).

2.3 Universalism, Globalization and Human Rights: A Theoretical Battlefield

One of the most important questions that has been raised since the adoption of the UDBHR and even during its development is whether it is possible to draft a text that could be applied across the world. Some argue that universalism is often seen as a western concept and others argue that globalization is an economic concept which can be applied to all nations (Bagheri 2011; Gracia 2014; ten Have and Gordijn 2014).

The UDBHR has been the object of many publications both positive and negative, not only about its universality but also about its relationship to human rights. It has been claimed that, "On the whole, it can be stated that the inclusion of bioethical norms, into human rights norms has not resulted in the collision of such norms, nor has it enhanced the relativity of international law" (Sandor 2008). In a world where cultural relativism is not absent and where constructivism has questioned the foundations of everything called "a norm" or "a principle", it is not surprising that Declarations of this kind are sometimes seen as unproductive or even useless. But not everybody agrees with such a view. The numerous articles published since the adoption of the various bioethics Declarations, especially since the adoption of the

Universal Declaration on Bioethics and Human Rights, show that they are also seen as relevant. As Bagheri wrote about the UDBHR, "The fact that this declaration has attracted many experts in the field indicates that they assumed this document will have a significant impact on bioethics worldwide, as an academic discipline, as a social discourse in general and on bio-policy in particular" (Bagheri 2011).

2.4 An Action Oriented Approach

Certainly it is important to continue the discussion about these theories, but it is also important to take a look at what has been accomplished since the adoption of the UDBHR. As the first Director General of UNESCO Julian Huxley wrote about reconciliation between nations, "It can be approached from above and from outside, as an intellectual problem, a question of agreement in principle: and it can also be approached from below and from within, as a practical problem, a question of agreement through action. The world is potentially one, and human needs are the same in every part of it" (Huxley 1946).

Now, after 20 years of UNESCO's active involvement in setting international norms in bioethics, these are important questions to ask: Are the declarations useful? Are they contributing to the achievement of UNESCO's mission? Are they useful in the drafting of national legislation? Are they giving rise to training programs and to the development of bioethical infrastructures such as Bioethics research and clinical committees? Further, are they contributing to the development of bioethics as an important tool of critical thinking in ethics?

We have considered the four following steps to reflect on these questions relating to the usefulness of the declaration on practice: Development, Implementation, Promotion and Knowledge construction. Development includes all the preliminary steps that lead to the adoption of a declaration. Implementation could include setting up bioethics committees, training, capacity building and the development of legislation or guidelines. Promotion will happen through publications, brochures, books, use of the web to present what is happening in different countries following the adoption of the UDBHR. Finally, knowledge construction could include research, academic articles and books publications as well as conferences.

Let us, as an example, focus in this article on the usefulness of the Universal Declaration on Bioethics and Human Rights (UDBHR) and consider the four steps in the realization of that declaration; development, implementation, promotion and knowledge construction.

Development The proposition to develop a declaration comes from either the director General of UNESCO or the Member States. For instance the idea to draft a declaration on the human genome came from Federico Mayor, the one on human genetic data from Koïchiro Matsuura. The UDBHR was suggested at a meeting of science ministers in 1999 in Budapest. In their final *Declaration on Science and the Use of Scientific Knowledge*, preoccupied by the pace at which science and technologies were progressing and the importance of sharing knowledge with less developed countries, they wrote: "The full and free exercise of science, with its own values, should not be seen to conflict with the recognition of spiritual, cultural, philosophical and religious values; an open dialogue needs to be maintained with these value systems to facilitate mutual understanding. For the development of an all-encompassing debate on ethics in science, and a possibly ensuing code of universal values, it is necessary to recognize the many ethical frameworks in the civilizations around the world" (World Conference on Science 1999).

In 2001, the Director-General organized a round table of Ministers of Science on bioethics during the 31st General Conference. The ministers, in their final communiqué invited UNESCO to examine the possibility of developing a universal instrument on bioethics. The General Conference adopted a resolution inviting the Director-General to submit the "technical and legal studies regarding the possibility of elaborating universal norms on bioethics" (UNESCO 2001).

The feasibility study prepared by the IBC concluded that a Declaration containing general principles could be prepared and would be useful. This idea was also supported by the then President of France, Jacques Chirac who, following the failure of a proposition from France and Germany to ban human cloning that was rejected in New York, declared that a normative instrument should be prepared by UNESCO.

The steps that have to be followed to finalize a declaration are complex. Member States usually ask for a comprehensive report on the issues at stake to be able to assess the usefulness of the enterprise. Then an expert advisory committee is put in place to draft a text which is presented to a meeting of governmental experts and then to the General Conference that will approve the text, ask for more work to be done or reject it. In the case of the UDBHR, Member States asked the IBC to conduct extensive consultation with them, the scientific community and civil society. These consultations were conducted in many countries (Turkey, Lithuania, Mexico, Iran, etc.) to discuss the text with governments, researchers and non-government organizations to be able to take into consideration the different contexts in which such an instrument would be used. Many articles reflect the preoccupation of reaching the international community at large. Extraordinary sessions and online consultations were held. So, even though some critics wrote that the consultation process was not broad enough, it can be confirmed that consultations were pursued on a large scale. Certainly not everything that was suggested found its way into the final text but many important issues raised during the consultations were reflected in the final text.

Implementation A declaration without an implementation strategy can rapidly become useless. Each of the three UNESCO declarations contains articles describing the monitoring process. UNESCO has done some important work to help member states use the declarations. Ethics committee have been put in place and continue to be supported in 17 countries, and many other countries from different regions of the world have approached UNESCO to create similar structures (UNESCO 2013). On the implementation side, training materials like the Core Curriculum (UNESCO

2008), Study Materials (UNESCO 2011) and different casebooks as well as the booklet on the setting up of Bioethics Committees are all sound illustrations of a good implementation process. Training courses have been conducted in 10 countries and 11 UNESCO Chairs in Bioethics have been established (UNESCO 2013).

Promotion Article 25.1 of the UDBHR states that, "UNESCO shall promote and disseminate the principles set out in this Declaration. In doing so, UNESCO should seek the help and assistance of the Intergovernmental Bioethics Committee (IGBC) and the International Bioethics Committee (IBC)" (UNESCO 2005). Considering what has been achieved, it can be concluded that UNESCO has done some good work.

Upon reviewing the Global Ethics Observatory (GEObs 2014. Retrieved from: www.unesco.org/new/en/social-and-human-sciences/themes/global-ethics-observatory/, n.d.), a database launched by UNESCO in 2005, we can see that information and expertise sharing has grown since the adoption of the UDBHR. It now contains 1510 experts, 497 institutions, 235 education programs, 738 legal or regulatory instruments, 151 codes of conduct, 416 resources in bioethics and applied ethics in science and technologies (UNESCO 2013). To promote the dissemination and application of the UDBHR, the IBC has published reports, with the help of specialists, that explain the ethical implications of the principles. These reports include some case studies that contextualize the issues discussed (UNESCO 2013).

Knowledge Construction Declarations are drafted and approved at a set time in the evolution of the instrument. They are never perfect and this assertion applies equally to the UDBHR, but is like building blocks in the construction process of theory and action. This especially applies to the UDBHR which is based on an ongoing conversation about the understanding and growth of the concept of bioethics. As has been argued, "This is not only a matter of codifying legal norms, but also of a general maturing of ideas, which help to identify and to delineate the nature and scope of common issues confronting humanity at a given stage of its evolution" (UNESCO 2007b). It could have been decided to postpone the drafting of the UDBHR until a more common understanding of the principles could be achieved but the Member States in their wisdom felt that immediate action was required given the fast pace at which science and technologies were developing and the questionable practices being adopted in developing countries (practices such as conducting clinical trials without ethics review, collecting data without having obtained consent, publishing scientific articles without involving researchers from participating countries and so on). A universal declaration could certainly be challenged but such challenges would keep the debate alive by calling for the involvement of all interested or affected stakeholders, and thus assist Member States to develop their own legislation and regulations.

In this knowledge construction process, as already mentioned in this article, one of the most important discussions today is about globalization and universality. The question being: is it possible to have a global bioethics? Universality being still seen as a western concept and globalization, being a fairly new concept, is seen as a work in progress in the moral arena. UNESCO, being an organization often called the conscience of the United Nations and working with its Member States to find ways to live together in peace, has seen the challenges posed by the development of science, technologies, communication and information. Its Directors General have often called to mind the role of the organization in the pursuit of the common good of humanity as a whole. Koïchiro Matsuura, its former director said in 2000, "It is UNESCO's duty to sound the alarm about the dangers of globalization and constantly to recall the need for equality of access for all to what some call the 'common good' […] Globalization is today generating uncharted challenges calling for new norms or ethical principles – or even regulatory mechanisms – with which to guarantee the continued exercise of universally recognized human rights. Many, if not all these challenges, fall squarely within UNESCO's defined responsibilities (UNESCO 2000)."

In the drafting stage, the title proposed for the UDBHR was *Declaration on universal norms on Bioethics*. However, the IBC was preoccupied with the importance of taking into account cultural diversity and the need to contextualize the application of the principles, and instead suggested the title *Universal Declaration on Bioethics and Human Rights*, thus avoiding the problem of proposing "universal norms". This new title was accepted by Member States. This change in the title and article 26 are a clear indication of the possibility of taking into account different religious and cultural contexts in the implementation of the Declaration. Article 26 states, "This Declaration is to be understood as a whole and the principles are to be understood as complementary and interrelated. Each principle is to be considered in the context of the other principles, as appropriate and relevant in the circumstances" (UNESCO 2005).

Other articles reflect the pragmatic thinking of the IBC and the members of the Inter-governmental Bioethics Committee (IGBC), for example Article 1 on the Scope of the Declaration, Article 14 on social responsibility and health, Article 15 on the sharing of benefits and Article 19 on ethics committees that should call for ongoing debate, education, public awareness and engagement in bioethics (UNESCO 2005).

2.5 Conclusion

The members of the IBC were selected from 36 different countries with different, religious and socio-cultural backgrounds. There are 191 Member States who adopted the declaration. Through the discussions of the IBC, committee members were aware of the challenges posed by the implementation of the Declaration yet they were more than ready to take on the challenge of implementing the UDBHR in their countries "as appropriate and relevant in the circumstances" as stated by article 26.

The title of Article 16 of the UDBHR is "Protecting future generations". This calls for a program of work of the IBC that will be forward looking. The challenge will be to deal with social, scientific and cultural issues that are already facing us for instance: cultural diversity; power sharing between scientists, governments, civil society, corporations and all nations; global disasters (infectious diseases); end of life issues; new technologies; as well as academic research on the definition of global bioethics as a discipline and a praxis. Langlois wrote, "With regard to the usefulness of the UNESCO declaration [UDBHR], the significance of its adoption as the first intergovernmental instrument on bioethics must be matched by action in the form of capacity building for it to be of added value in the realm of biomedical research ethics" (Langlois 2008). To that statement we can confidently answer that the actions taken since the adoption of the UDBHR are a clear indication of the usefulness of this instrument especially in countries where no bioethics infrastructures were previously in place.

References

Bagheri, A. 2011. The impact of the UNESCO declaration in Asian and global bioethics. *Asian Bioethics Review* 3(2): 52.
Global Ethics Observatory (GEobs). Available at: http://www.unesco.org/new/en/social-and-human-sciences/themes/global-ethics-observatory/. Last visited 16 Jan 2014.
Gracia, D. 2014. History of global bioethics. In *Handbook of global bioethics*, 19–34. Dordrecht: Springer.
Huxley, J. 1946. *UNESCO its purpose and its philosophy*. Preparatory Commission of the United Nations Educational, Scientific and Cultural Organization. Available at: http://unesdoc.unesco.org/images/0006/000681/068197eo.pdf. Last visited 16 Jan 2014.
Langlois, A. 2008. The UNESCO universal declaration on bioethics and human rights: Perspectives from Kenya and South Africa. *Health Care Analysis* 16: 19.
Sandor, J. 2008. Human rights and bioethics: Competitors or allies? The role of international law in shaping the contours of a new discipline. *Medicine and Law* 27: 15–28.
Ten Have, H., and B. Gordijn. 2014. Introduction. In *Handbook of global bioethics*, 3–18. Dordrecht: Springer.
UNESCO. 1945. Constitution. Available at: http://portal.unesco.org/en/ev.php-URL_ID=15244&URL_DO=DO_TOPIC&URL_SECTION=201.html. Last visited 19 Jul 2013.
UNESCO. 2000. Director General advocates access for all to "common good", Paris, October 9. Available at: http://mailman.apnic.net/mailing-lists/s-asia-it/archive/2000/10/msg00007.html. Last visited 12 Jan 2014. It is worth noting that there are 121. 000 mentions on Google about UNESCO and the common good.
UNESCO. 2001. 31/*Resolution 22*. 31st Session of the General Conference of UNESCO. Available at: http://unesdoc.unesco.org/images/0012/001246/124687f.pdf. Last visited 15 Jan 2014. Also referenced in: Ten Have, H.A.M.J. and Michèle S. Jean (eds.). 2009. *Universal declaration on bioethics and human rights. Background, principles and application*. Paris: UNESCO Publishing.
UNESCO. 2005. Available at: http://portal.unesco.org/en/ev.php-URL_ID=31058&URL_DO=DO_TOPIC&URL_SECTION=201.html. Last visited 12 Jan 2014.
UNESCO. 2007a. *Standard-setting in UNESCO*, Normative action in education, science and culture, vol. 1, ed. Abdulqawi, 12. Paris: UNESCO.
UNESCO. 2007b. *Standard-setting in UNESCO*, Normative action in education, science and culture, vol. 1, ed. Abdulqawi, 13. Paris: UNESCO.
UNESCO. 2008. Bioethics core curriculum, Section 1: Syllabus ethics education program, Paris. Available at: http://unesdoc.unesco.org/images/0016/001636/163613e.pdf. Last visited 12 Jan 2014.

UNESCO. 2011. Study materials, Section 2: Ethics education program, Paris. Available at: http://unesdoc.unesco.org/images/0021/002109/210933e.pdf. Last visited 12 Jan 2014.
UNESCO. 2012. *Basic texts*, 2012th ed, 117–119. Paris: UNESCO.
UNESCO. 2013. 1993–2013: 20 years of bioethics at UNESCO. Available at: http://unesdoc.unesco.org/images/0022/002208/220865e.pdf. Last visited 15 Jan 2014.
World Conference on Science. 1999. Par.32. Available at: http://unesdoc.unesco.org/images/0012/001229/122938eo.pdf. Last visited 15 Jan 2014.

Chapter 3
Twenty Years of the International Bioethics Committee: Achievements and Future Priorities

Nouzha Guessous

Abstract In the 20 years since the establishment of the International Bioethics Committee (IBC), UNESCO has become a key interlocutor on bioethics through its three declarations and the Ethics Education Programme, all of which bear the hallmarks of a multidisciplinary and pluralist approach that seeks to balance universal and contextual considerations. Over the next 20 years, the IBC must continue to ensure that scientific and technological advances do not exacerbate human vulnerability, particularly in resource-poor countries. Issues such as the trafficking of human organs and tissue, the migration of health workers, and the dangers of counterfeit medicines should be considered with a view to making practical recommendations. Support for bioethics committees and bioethics education in developing countries must remain a priority, and the governments of Member States must be involved in this process to ensure its sustainability.

3.1 Leadership of UNESCO and the International Bioethics Committee in Global Bioethics

The reflections and comments in this section are informed by my experience of working with the IBC, initially as a member (2000–2007) and then as the IBC Chairperson (2005–2007). Now, 20 years after the establishment of the IBC, it is clear that UNESCO as an international and intergovernmental organization is a key interlocutor on bioethics issues, for its Member States as well as for the global community. UNESCO has put bioethical issues on the agenda of the United Nations and

---

An early draft of this article was presented in the UNESCO Symposium, "The role of UNESCO in bioethics for the next 20 years symposium", in Paris on September 6, 2013.

N. Guessous (✉)
Hassan II University of Casablanca and Centre Jacques Berque,
Casablanca & Rabat, Morocco
e-mail: nouzhaguessous@gmail.com

© Springer International Publishing Switzerland 2016
A. Bagheri et al. (eds.), *Global Bioethics: The Impact of the UNESCO International Bioethics Committee*, Advancing Global Bioethics 5,
DOI 10.1007/978-3-319-22650-7_3

international governance. By approaching bioethics as a discipline and praxis, UNESCO has promoted standard-setting actions and capacity building in Member States, drawing on the expertise of the IBC members and under the leadership of the Division of Ethics of Science and Technology. As part of this process, the ethics program aimed to build a bridge between decision-makers and legislators on the one hand, and researchers in science and technology on the other. Through training and awareness-raising measures, UNESCO has helped Member States to open up the debate between scientists and decision-makers to include the general public. This vision of IBC leadership is founded on:

1. Three declarations, namely the *Universal Declaration on the Human Genome and Human Rights* (UNESCO 1997), the *International Declaration on Human Genetic Data* (UNESCO 2003a), and the *Universal Declaration on Bioethics and Human Rights* (UNESCO 2005). These declarations were the result of consultations between scientists, those working in the field of bioethics, independent IBC experts, members of the Intergovernmental Bioethics Committee (IGBC) and government experts, as well as other United Nations agencies with an interest in bioethics notably the World Health Organization (WHO). Today, these three declarations provide an international legal and moral framework for all Member States. The reports and recommendations drafted and adopted by the IBC complement and clarify this standard-setting framework by developing regulations at the national level.
2. The Ethics Education Program of the Division of Ethics of Science and Technology, including the Assisting Bioethics Committees (ABC) program, which has promoted and assisted the formation of ethics committees in many developing countries; the establishment of the Global Ethics Observatory (GEObs), and the creation in 2008 of a bioethics training module for medical students (Ten Have 2006). A core course in bioethics aimed primarily at medical students can be found online in all the working languages of UNESCO (2008).
3. The initiative for the establishment of the United Nations Inter-Agency Committee on Bioethics to coordinate action in the field of bioethics across the United Nations system.

3.2 The IBC: A Pluralist and Multidisciplinary Setting

One of the defining characteristics of the success of all these initiatives is the overarching multidisciplinary and pluralist approach of the IBC. This approach seeks to balance the universal against the particular, through consensus whenever possible, or if consensus seems impossible or reductive, by synergizing efforts to present different points of view in such a way that they become fully comprehensible to all.

I wish to emphasize that this was the outstanding feature of my participation in the IBC. Engaged interaction between the members of the Committee with such diverse backgrounds, training, cultures, and experiences is a key to the success of this committee in bioethical discourse. It should be noted that the IBC was pluralist in its approach even before that concept became part of Article 12 of the Universal Declaration on Bioethics and Human Rights.[1] The fruits of this pluralism include the Committee's reports on stem cells (UNESCO 2001) and pre-implantation genetic diagnosis (UNESCO 2003b). Discussion within the committee is characterized by respectful listening of all different opinions to enable the IBC members to find consensus and/or compromises that are acceptable to the members of the IBC initially and to UNESCO Member States through the Intergovernmental Bioethics Committee.

Idealism aside, it can be claimed from that experience that if such an ethical approach could prevail when discussing the major and minor issues facing humanity, we would have far fewer conflicts and wars, and that the pluralism of the IBC approach and debate gave expression to the concept of respect for cultural diversity as enshrined in Article 12 of the Declaration.

3.3 Challenges for the Next 20 Years

At the organizational level, in order to make the best use of resources, avoid duplication of efforts and an overlap of responsibilities, it is important to review coordination between UNESCO bodies (IBC, IGBC, and World Commission on the Ethics of Scientific Knowledge and Technology (COMEST)) and other agencies of the United Nations system, particularly the World Health Organization and the Inter-Agency Committee.

The recommendations of the 20th session of the IBC, and its vision of the road map for the next 20 years, should be adhered to with regard to priority issues and actions. In particular, the recommendation that bioethics issues should always be addressed in the context of human rights, justice, and respect for human dignity must remain a central tenet. By following this tenet, UNESCO will ensure that it acts as the ethical conscience of the United Nations when pursuing all programs and initiatives. One prominent example of a priority issue is the risk to vulnerability posed by new biomedical technologies. In fact, more global attention is required on the subject of new risks threatening vulnerable individuals and groups as new biomedical technologies are developed. However, the frequency, gravity and diversity of these risks are increasing at an alarming rate, as illustrated by the IBC's call to highlight these new risks (UNESCO 2013).

It is therefore absolutely vital that UNESCO, together with other relevant agencies of the United Nations system, intensifies efforts to ensure that advances in science and technology are not used to exploit and aggravate such vulnerabilities. As

---

[1] Article 12 of the *Universal Declaration on Bioethics and Human Rights*: "The importance of cultural diversity and pluralism should be given due regard. However, such considerations are not to be invoked to infringe upon human dignity, human rights and fundamental freedoms, nor upon the principles set out in this Declaration, nor to limit their scope".

emphasized by the IBC at its most recent session, ethical governance at all levels, including at the technological and scientific level is central to the question of how to protect vulnerable individuals and groups.

The crux of the problem is corruption, which in all its forms and at all levels spreads like a cancer, invading every organ of the human family and preventing the creation of "equitable societies" that have the potential to be sustainable. Currently, one of the problems perpetuated by corruption is counterfeit medicines. Various studies have shown that 10 % of all medicines in global circulation are counterfeit, with that figure rising as high as 20 % or 30 % of the market in some regions of South America, Asia, and above all Africa. WHO estimated that the market for counterfeit medicine, which is even more lucrative than that for illicit drugs, worth $75 billion in 2010 (WHO 2006). The problem is exacerbated and perpetuated by the lack of pharmaceutical regulation and control. This can be demonstrated by looking at countries such as Australia, Canada, Japan, New Zealand, the United States of America and most European countries, where the incidence of counterfeit medicines is less than 1 % (WHO 2012). This takes us to the heart of the issue of ethical governance at national and local levels, and the equation (corruption + poverty) = (production + aggravation of vulnerability). Ultimately, the market in counterfeit medicine targets and exploits vulnerable poor people who cannot afford the current excessive cost of medicines manufactured to international standards of quality and safety. It is vitally urgent that UNESCO examine this issue in partnership with other United Nations system agencies.

3.4 Priorities for the Future

Following list suggests my opinion on the priorities in UNESCO bioethics programs.

(1) Promote, protect and strengthen human rights in and through progress in science and technology. This means:

   1.1. Completing work on Article 8 of the Universal Declaration on Bioethics and Human Rights[2] promoting the principle of respect for human vulnerability and personal integrity in view of the increasing threat of exploitation of vulnerability. Specific answers must be found to the following questions:

   – How can we protect vulnerable individuals and groups in view of the increasingly serious and diverse risks of the commodification of the

---

[2]Article 8 of the 2005 *Universal Declaration on Bioethics and Human Rights*: "In applying and advancing scientific knowledge, medical practice and associated technologies, human vulnerability should be taken into account. Individuals and groups of special vulnerability should be protected and the personal integrity of such individuals respected."

human body at the international level, such as the sale of organs and tissue?
- Current debates on the regulations that are in force or are in the process of coming into force regarding gestational surrogacy appear to be moving towards an idealistic discourse. Where is the line to be drawn between altruism and the exploitation of vulnerability? Women remain vulnerable in many regions of the world and gestational surrogacy is an additional risk in terms of the commodification of their bodies. This could be a reality for hundreds of millions of women worldwide who are living in extreme poverty and deprivation, and have no alternative but to "rent out" their wombs to others on the basis of need rather than altruism and generosity.

1.2. To continue reflective work and the drafting of reports and recommendations on the implementation and promotion of the principles of the Universal Declaration on Bioethics and Human Rights, particularly Article 12 on the principle of respect for cultural diversity and pluralism. How can we promote intercultural dialogue while respecting the cultural context and the universal principles of bioethics and human rights?

1.3. Lastly, the issue of the migration of healthcare workers is a matter of great concern. In 2006, WHO estimated the shortfall in the number of healthcare professionals worldwide at 4.3 million. Low-income countries are particularly badly affected, and of the 57 countries where the shortage was deemed to be critical, 36 were in sub-Saharan Africa (OECD 2010). We must therefore consider this issue – which is another form of commodification of human beings to the detriment of the principles of justice, responsibility and solidarity – in order to propose legally binding international solutions.

(2) Practical action should target developing countries as a priority.
This means:

2.1. Continue to promote and assist national bioethics committees with the involvement and engagement of the governments of the countries concerned to ensure sustainability and independence.

2.2. Promote and support teaching and education on bioethics for all professionals and decision-makers in the fields of health and research.

As highlighted by the IBC at its most recent session, these priorities could be used as a barometer for UNESCO's achievements and those of other United Nations system agencies in the field of bioethics.

References

OECD. 2010. International migration of health workers. Improving international cooperation to address the global health workforce crisis. Available at: http://www.oecd.org/migration/mig/44783473.pdf. Last visited 26 Mar 2014.

Ten Have, H. 2006. The activities of UNESCO in the area of ethics. *Kennedy Institute of Ethics Journal* 16(4): 333–351. The Johns Hopkins University Press, http://www.unesco.org/new/fileadmin/MULTIMEDIA/HQ/SHS/pdf/KIEJ-2006.pdf

UNESCO. 1997. Universal Declaration on the Human Genome and Human Rights. Available at: http://portal.unesco.org/en/ev.php-URL_ID=13177&URL_DO=DO_TOPIC&URL_SECTION=201.html. Last visited 26 Mar 2014.

UNESCO. 2001. IBC report on the use of embryonic stem cells in therapeutic research. Available at: http://unesdoc.unesco.org/images/0013/001322/132287e.pdf. Last visited 26 Mar 2014.

UNESCO. 2003a. International Declaration on Human Genetic Data. Available at: http://portal.unesco.org/en/ev.php-URL_ID=17720&URL_DO=DO_TOPIC&URL_SECTION=201.html. Last visited 26 Mar 2014.

UNESCO. 2003b. Report of the Working Group of the IBC on pre-implantation genetic diagnosis and germ-line interventions. Available at: http://unesdoc.unesco.org/images/0013/001302/130248e.pdf. Last visited 26 Mar 2014.

UNESCO. 2005. Universal Declaration on Bioethics and Human Rights. Available at: http://portal.unesco.org/en/ev.php-URL_ID=31058&URL_DO=DO_TOPIC&URL_SECTION=201.html. Last visited 26 Mar 2014.

UNESCO. 2008. Core course in bioethics. Available at: http://unesdoc.unesco.org/images/0016/001636/163613f.pdf. Last visited 26 Mar 2014.

UNESCO. 2013. Scientific progresses and new risks of discrimination. A call of the International Bioethics Committee. Available at: http://www.youtube.com/watch?v=rc6UH8x_Chw&feature=player_embedded. Last visited 26 Mar 2014.

WHO. 2006. Counterfeit medicines: The silent epidemic. Available at: http://www.who.int/mediacentre/news/releases/2006/pr09/en/. Last visited 26 Mar 2014.

WHO. 2012. Medicines: Spurious/falsely labelled/falsified/counterfeit (SFFC) medicines. Fact sheet no. 275. Available at: http://www.who.int/mediacentre/factsheets/fs275/en/. Last visited 26 Mar 2014.

Chapter 4
The Universality of the UNESCO *Universal Declaration on Bioethics and Human Rights*

Richard Magnus

Abstract This chapter discusses the acceptance of the UNESCO *Universal Declaration on Bioethics and Human Rights* (UDBHR), by acclamation. It highlights that the Declaration has broadened the scope of bioethics, by integrating international human rights law into the field of biomedicine, and considered about the environment biosphere and biodiversity. The Declaration is seen as a global benchmark for all countries, as it focuses on the concepts of common principles, shared values and internal cooperation between countries. This paper also critically analyses the limitations of the UDBHR, commenting on the difficulty in the application of human rights and solidarity in the Asian and global context, respectively. However, despite its limitations, the Declaration is still relevant and applicable for all, as it takes into consideration the diverse culture in different countries. The UDBHR also strives to offer a more comprehensive understanding of the principles, by issuing further guidance.

4.1 Introduction

In 2005, the General Assembly of the United Nations Educational, Scientific and Cultural Organization (UNESCO) adopted by acclamation the *Universal Declaration on Bioethics and Human Rights* (UDBHR). Bioethics had traditionally focused on the relationship between healthcare providers and their patients, but rapid advancement in biotechnology led to a global consensus that scientific and technological progress must be accompanied by careful consideration of ethical and social issues that may arise. Following a feasibility study to assess the possibility of elaborating an international instrument on bioethics, the 32nd UNESCO General Conference considered it "opportune and desirable [for UNESCO] to set universal standards in the field of bioethics for due regards for human dignity and human

R. Magnus (✉)
Chairman, Bioethics Advisory Committee, Singapore

Member, UNESCO International Bioethics Committee, UNESCO, Paris, France
e-mail: rmagnus@singnet.com.sg

rights and freedoms, in the spirit of cultural pluralism inherent in bioethics" (UNESCO 2003). It was with the aim of providing ethics guidance for biomedical advancement that the UNESCO's International Bioethics Committee (IBC) started to develop the UDBHR in 2003.

In celebration of the 20th anniversary of the IBC, this chapter considers the universality of the Committee's most ambitious achievement to date. The UDBHR was the first international instrument to establish global standards on bioethics, the scope of which was broadened to include life sciences, in addition to the field of medicine. The Declaration also shifted the emphasis of bioethics from individuals to wider society and humankind in general, by focusing not just on personal autonomy, but on respect for human dignity and human rights as well. As the UDBHR focuses on basic principles and leaves the implementation details to the nation states, it is generally acceptable and easily applicable for developed and developing countries.

4.2 Acceptance of the Declaration by Acclamation

> "A body of independent experts (the IBC) drafts the instrument with nonbinding advice and comment from government-appointed officials (the IGBC). The draft is then subject to negotiation and revision by government-appointed policy experts, after which it is ultimately submitted for acceptance or rejection by the General conference itself. Throughout the process, the procedural standard of acceptance is *consensus*." (Snead 2009) [Emphasis added].

The UDBHR was the first international effort for a global approach to bioethics. Its preparation involved both governmental and non-governmental actors; with expertise in various disciplines such as bioethics, law, life sciences, and social sciences; and from as many as 191 member states of UNESCO. During the drafting stage, the IBC consulted widely, engaging with various stakeholders from and on multiple levels. On the global platform, through an Interagency Committee on Bioethics established by the UN Secretary-General, UNESCO led discussions on matters of common concern with other international bodies like the World Health Organization, World Trade Organization, and Food and Agriculture Organization. Regional experts were consulted, as were national bioethics experts from several countries. In addition, the IBC also engaged in dialogue with representatives of civil society organisations, different religious bodies and traditions, scientists and other experts through a major public symposium (Kirby 2008 and 2009). The result of such a multi-national, multi-cultural, multi-faceted collaborative effort was the identification of 15 basic principles that provide normative guidance not just for member states, but also for "individuals, groups, communities, institutions and corporations, public and private" (Article 2(b)). UNESCO achieved its aim of receiving "*the broadest acceptance possible* by public authorities, the scientific community and general public" (UNESCO 2003, p. 10), when the Declaration was adopted "unanimously, without any contrary votes or recorded abstentions" (Kirby 2008 and 2009). That all 191 member states endorsed the Declaration despite different religious and socio-cultural backgrounds, showed that UNESCO had managed to find some common grounds between nations, in order to set minimum standards for bioethics that are universally acceptable.

4.3 Broadening the Scope of Bioethics

The first principle advocated in the UDBHR is respect for "human dignity, human rights and fundamental freedoms" (Article 3). Although this principle had long been established since the 1948 *Universal Declaration of Human Rights* (United Nations 1948), which is today widely regarded as the cornerstone of the international human rights system that emerged after the Second World War, the UDBHR was the first international instrument to comprehensively integrate international human rights law into the field of biomedicine. By broadening the scope of the respect principle from personal autonomy to human dignity, the UDBHR overcomes a shortcoming of previous bioethics documents, which seemed to accord respect only to autonomous persons. The UDBHR on the other hand, "includes the protection of those who are not yet, or are no more, morally autonomous (new-born infants, senile elderly, people with serious mental disorders, comatose patients, etc.)" (Andorno 2009). In addition to existing persons, the UDBHR also covers "future persons" as respect for human dignity requires that new challenging practices, such as reproductive cloning or germ-line interventions, do not result in the modification of basic human features or affect the integrity of the human species (Andorno 2009). The UDBHR does not precisely define "human dignity", but one may gain a better understanding of the concept from the *Universal Declaration of Human Rights*, which recognised the "inherent dignity … and equal and inalienable rights of all members of the human family" (Preamble). It emphasises that dignity is an unconditional worth that every human being has regardless of their intellectual or moral abilities. Thus, every human being is entitled to fundamental rights and freedom, and they should be protected from any harm to their dignity even though they have consented to such acts.

Integrating the human rights framework into bioethics is meaningful because most basic human prerogatives are relevant to biomedical activities, for instance the rights to life, physical integrity, privacy and access to basic health care (Andorno 2007, 2009). There are fundamental rights and freedoms that are desirable to all human beings, regardless of ethnicity, gender, age, nationality or socio-economic status. The UDBHR identified some of these entitlements – for surely all persons must wish to be treated with due respect (Articles 3, 5, 8, and 9), be treated justly and equitably (Article 10), to enjoy the highest attainable standard of health (Article 14), and not be subject to discrimination and stigmatisation (Article 11). Hence, just like the *Universal Declaration of Human Rights*, the UDBHR is a proclamation of "the highest aspiration of the common people". The UDBHR has also broadened the scope of bioethics to include considerations about the environment, biosphere and biodiversity (Article 17). It is no longer possible to advance science and technology without reflecting on the impact of our actions on our environment and other living beings. Hence, the UDBHR is indeed a universal instrument, for in addition to encouraging consideration for our fellow human beings, it also recognises the need to give due regard to other living beings that we share this earth with.

4.4 Global Benchmarks for All

By focusing on common principles and shared values, the UDBHR established global standards that could be used by both developing and developed countries. As many of the UNESCO member states were lacking a well-developed bioethics infrastructure, the UDBHR was intended to address this gap by setting "universal ethical benchmarks" (Wolinsky 2006; ten Have 2006).

A significant contribution of the UDBHR is the provision of guidance for transnational research (Article 21). The Declaration defined a set of minimum standards that all countries keen to conduct ethical scientific research could commit to, and outlined the roles and responsibilities of countries involved in transnational research. As an intergovernmental third party, UNESCO is in a unique position to ensure that while the international progress of science and technology are not unduly impeded, research participants worldwide are also properly protected. Hence, there is an underlying principle of global justice in the relevant articles; for example, Article 15(2) states that "[b]enefits should not constitute improper inducement to participate in research". As more clinical trials are being conducted in developing rather than developed countries due to various reasons, such guidance will ensure that vulnerable groups are protected from excessive harm and exploitation. The UDBHR also calls for appropriate action on global issues such as global health, bioterrorism and illicit trafficking of human biological materials. UNESCO plays a crucial part in the promotion of science and technology internationally for the benefit of all of humankind, and the UDBHR is a commendable effort towards that purpose. What sets UDBHR apart as a truly universal instrument on bioethics is its promotion of the principle of solidarity and internal cooperation (Articles 13 and 24), which is further supported by its other principles of social responsibility (Article 14) and sharing of benefits (Article 15). Unlike traditional bioethics discourse, which tends to place the emphasis mainly on individual persons, the UDBHR also focuses on larger society and even the global community. The Declaration highlights the role that all sectors of society can play in ensuring the ethical conduct of biomedical research and clinical practice. Indeed, it is even stated that the UDBHR is intended "to guide the actions of individuals, groups, communities, institutions and corporations, public and private" (Article 2-b). The UDBHR calls on the principle of solidarity to unite individuals and states in achieving global goals, particularly the promotion of health and social development (Article 14-1). By identifying the protection of health and social development as a "central purpose of governments", the goal is no longer mere moral persuasion but the subject of deliberate governmental policy (UNESCO *Report on Social Responsibility and Health* 2010). UNESCO member states that have adopted the Declaration have the obligation to protect citizens from health threats and social threats such as poverty or severe environmental degradation. Thus, bioethics is no longer just an academic field, as policy makers too are drawn into examining the issues related to medicine and life sciences.

Since "the enjoyment of the highest attainable standard of health is one of the fundamental rights of every human being without distinction of race, religion, political belief, economic or social condition" (Article 14-2), health should not only be confined to national interests and responsibilities. Instead, social relationships beyond state boundaries, such as international trade and transnational research activities should be encouraged (UNESCO *Report on Social Responsibility and Health* 2010). International civil society organisations like UNESCO play an important role in overcoming geographical boundaries and promoting universal access to affordable, essential medicines for *all* human beings. In fact, the UDBHR urges states to "respect and promote solidarity between and among States, as well as individuals, families, groups and communities, with special regard for those rendered vulnerable by disease or disability or other personal, societal or environmental conditions and those with the most limited resources"(Article 24-3). It is also on the basis of solidarity that benefits obtained from scientific research and its applications should be shared with wider society and the international community, particularly with developing countries (Article 15). The benefits identified in the UDBHR – such as provision of new diagnostic and therapeutic modalities or products stemming from research, access to scientific and technological knowledge, and capacity-building facilities for research purposes – will reduce the time taken to discover prevention measures and treatments for diseases as well as reduce the number of people to be involved in clinical trials. For example, benefit sharing will help many Asian countries cope with curable infectious diseases that they are still struggling with, as well as the new series of infectious diseases such as Avian Flu and SARS that have hit the continent (Bagheri 2011). In today's globalised world, where air travel makes it easy for people (together with the diseases that they may carry) to move across continents quickly, it is clearly advantageous for international cooperation in order to prevent transnational transmission. Therefore, infectious diseases are a global problem, the eradication of which is not exclusive to developing countries only. Furthermore, by sharing benefits, for instance in the form of resources, the number of clinical trials that need to be conducted will be reduced. This would lead to a decrease in the number of people needed for trials, which in turn will help to reduce the rate of exposure to harm and exploitation. Indeed, benefit sharing is necessary to attain the universal aim of good health for all of humankind.

Underlying the principle of solidarity is a sense of "connectedness", and it implies at the very least "the willingness to take the perspective of others seriously", which Gunson has suggested is a kind of "weak solidarity" that is more suited to inclusion in universal documents such as UDBHR (Gunson 2009). This is distinct from "strong solidarity", as it merely requires that one listen and assess whether a particular cause is worthy of allegiance and does not require action in support of specific goals or political causes. Strong solidarity simply would not be feasible in universal instruments because it would require that all nations involved actually be supportive of *all* possible projects. Nevertheless, the principle of solidarity, whether weak or strong, implies unity and a sense of belonging; and as advocated in the UDBHR, it emphasises "the interconnectedness of *humanity*" (Gunson 2009), a kind of global solidarity among *all* human beings.

4.5 Relevance of the Principles of the Declaration

The UDBHR focuses on basic principles of bioethics that have, and are intended to, withstand the test of time. In fact, several of the principles are derived from other existing bioethics documents.

The reference to human dignity and human rights is not an entirely new approach in bioethics (Andorno 2007). As Andorno pointed out, the 1997 *Convention on Human Rights and Biomedicine* by the Council of Europe requires that its member states "shall protect the dignity and identity of all human beings and guarantee everyone, without discrimination, respect for their integrity and other rights and fundamental freedoms" (Article 1). The World Medical Association's *Declaration of Helsinki on Research Involving Human Subjects* (1964, revised 2013) makes reference to the rights of participants, and promotes the protection of human dignity of research subjects, as well as respect for their life, health and privacy. The UN Commission on Human Rights Resolution 2003/69 entitled *Human rights and bioethics* stresses the need to ensure the protection of human rights in bioethics, and repeatedly mentions the "dignity of the human being". The ethics committee of the Human Genome Organisation has also issued several statements in which it encourages the need to "adhere to international norms of human rights" and to accept and uphold "human dignity and freedom". Nevertheless, the UDBHR remains as the first international instrument to take such a comprehensive approach to integrate respect for human dignity and human rights into the area of bioethics.

The requirement that individuals give free, prior and informed consent before participating in any medical intervention or scientific research, has been a key feature of bioethics since it was first introduced in the 1949 (Nuremberg Code 1949). The principle of informed consent, which is based on the principle of autonomy (Article 5) and is an affirmation of human rights and respect for human dignity (Article 3), has been espoused by intergovernmental organisations, and international bodies and documents. Indeed, Article 6(2) of the UDBHR captures the essential ideas on informed consent as encapsulated in another important bioethics document, the *Declaration of Helsinki*.[1]

---

[1] *UDBHR* Article 6 (2): "Scientific research should only be carried out with the prior, free, express and informed consent of the person concerned. The information should be adequate, provided in a comprehensible form and should include modalities for withdrawal of consent. Consent may be withdrawn by the person concerned at any time and for any reason without any disadvantage or prejudice."

Another principle that is not original to the UDBHR is the notion of maximising benefits and minimizing possible harms (Article 4), which was first articulated in the 1979 *Belmont Report* as one of two general rules formulated under the principle of beneficence. Generally, there is a global consensus that it is reasonable to expose research participants to some level of risks, provided an ethics committee or institutional review board determined that it is justified in light of potential benefits. These benefits may be direct to the research participants; or indirect through the development of generalised knowledge for the society, which could lead to, for example, an improved health care system. It is an obligation for the investigators and institutional review boards to give forethought to the maximisation of benefits and reduction of risks that might result from the study. These expected risks and benefits should then be made known to the potential research participants during the consent process. The need to protect vulnerable persons, highlighted in Article 8 of UDBHR, had been articulated earlier, namely in the 1979 *Belmont Report*, and subsequently in the 2002 *International Ethical Guidelines for Biomedical Research Involving Human Subjects of the Council for International Organizations of Medical Sciences*. As elaborated in the recently published UNESCO IBC's Report on *The Principle of Respect for Human Vulnerability and Personal Integrity*, "[t]he human condition implies vulnerability. Every human being is exposed to the permanent risk of suffering "wounds" to their physical and mental integrity. Vulnerability is an inescapable dimension of the life of individuals and the shaping of human relationships. To take into account human vulnerability acknowledges that we all may lack at some point the ability or the means to protect ourselves, our health and our well-being. We are all confronted with the possibility of disease, disability and environmental risks. At the same time, we live with the possibility that harm, even death, can be caused by other human beings" (UNESCO International Bioethics Committee 2013). Furthermore, "some groups of people can always be considerable vulnerable because of their status (e.g. children), [while] others may be vulnerable in one situation but not in another" (UNESCO International Bioethics Committee 2013, p. 13). Since there will always be individuals and groups of special vulnerability all over the world, for as long as the human species is alive, the need to protect vulnerable persons and respect their personal integrity is a universal principle, that ought to be observed throughout time.

Declaration of Helsinki: 26. In medical research involving human subjects capable of giving informed consent, each potential subject must be adequately informed of the aims, methods, sources of funding, any possible conflicts of interest, institutional affiliations of the researcher, the anticipated benefits and potential risks of the study and the discomfort it may entail, post-study provisions and any other relevant aspects of the study. The potential subject must be informed of the right to refuse to participate in the study or to withdraw consent to participate at any time without reprisal. Special attention should be given to the specific information needs of individual potential subjects as well as to the methods used to deliver the information.

After ensuring that the potential subject has understood the information, the physician or another appropriately qualified individual must then seek the potential subject's freely-given informed consent, preferably in writing. If the consent cannot be expressed in writing, the non-written consent must be formally documented and witnessed.

The need to protect future generations (Article 16) and the environment, biosphere and biodiversity (Article 17) are also principles that ought to be observed by humankind always, to ensure the continued survival, and betterment, of our species and the world we live in. The list of principles mentioned in this section is not exhaustive, and the examples given are meant to illustrate the enduring quality of the UDBHR. Certainly, societal values do change over time and ethics is fluid; but the UDBHR has succeeded in identifying certain universally shared values and principles, which is likely, and ought to, withstand the test of time.

4.6 Applicability of the Declaration

The basic terms and principles, such as 'bioethics', 'human dignity', 'solidarity', and 'justice', are not explicitly defined in the UDBHR. As Andorno explained, "[t]he same happens with all basic moral and legal principles (justice, freedom, autonomy, etc.)…not only because of the impossibility of finding a precise definition of such fundamental concepts that satisfies everyone, especially in a transcultural context, but also because lawmakers are well aware that rigid definitions may in some cases lead to unsolvable difficulties in implementation of legal norms" (Andorno 2009). By leaving out the exact definition of these key concepts, the UDBHR is relevant, and can be applied by all states, despite divergent interpretations. As Gunson poignantly notes, it is a delicate business trying to strike a workable balance between specificity and normativity versus universality and consensus (Gunson 2009).

The Article 12 of the UDBHR calls on respecting cultural diversity and pluralism. This recognition of possible divergent positions on sensitive issues =as many bioethical questions are- greatly contributed to the easy acceptance of the UDBHR. In today's globalised world, it would no longer be possible to ignore (or be intolerant of) multiculturalism and pluralism of beliefs and viewpoints. For example, the Declaration places great importance on respecting individual autonomy (Article 5), but it also notes that difficulties may arise when implementing the principle in situations where communal forms of decision making may be prevalent (UNESCO International Bioethics Committee 2008). To accommodate both the cultural requirements as well as to protect the rights and interests of the person, the Declaration allows for seeking the additional agreement of legal representatives or community, but "[i]n no case should a collective community agreement or the consent of a community leader or other authority substitute for an individual's informed consent" (Article 6-3). By making provisions for diversity, the Declaration essentially becomes applicable for most, if not all countries.[2]

---

[2] Article 12 further states that "such considerations are not to be invoked to infringe upon human dignity, human rights and fundamental freedoms, nor upon the principles set out in this Declaration, nor to limit their scope". This may pose constraints on nations that are not in adherence with international human rights law.

In addition, the UDBHR also recognises the need to give due regard to local laws. The Declaration encourages consideration of domestic law, particularly in relation to consent (Articles 6 and 7). While "[s]cientific research should only be carried out with the prior, free, express and informed consent of the person concerned…[e]xceptions to this principle should be made only in accordance with ethical and legal standards adopted by States, consistent with the principles and provisions set out in this Declaration, … and international human rights law" (Article 6-2). Although it is ideal to obtain informed consent *prior* to one's participation in research or receiving treatment, there are circumstances, for example during emergency situations, where it may be impossible to do so due to the person being unconscious or confused. UNESCO is clearly sensitive to the fact that an instrument of international status cannot be too proscriptive and has therefore included provisions for exceptions. At the same time, appropriate safeguards have also been built into the UDBHR to protect against abuse. Though cultural diversity and pluralism should be respected, such considerations do not justify infringement of human dignity, human rights and fundamental freedoms, nor upon the UDBHR principles, nor to limit their scope (Article 12). Article 27 also states that "If the application of the principles of this Declaration is to be limited, it should be by law, including laws in the interests of public safety, for the investigation, detection and prosecution of criminal offences, for the protection of public health or for the protection of the rights and freedoms of others. Any such law needs to be consistent with international human rights law".

Specifically, in the context of research involving persons without the capacity to consent, UDBHR requires that such research should normally be conducted if there are direct health benefits to the person; but if not, the study should only be "undertaken by way of exception, with the utmost restraint, exposing the person only to a minimal risk and minimal burden and, if the research is expected to contribute to the health benefit of other persons in the same category, subject to the conditions prescribed by law and compatible with the protection of the individual's human rights" (Article 7 -b).

4.7 Limitations in the Universality of the Declaration

While the UDBHR acknowledges the existence of cultural diversity, it has been argued that it fails to uphold its "validity". With repeated mention on human dignity and human rights in the Declaration, it seems to place limitations for the applications of other values (Hayry and Takala 2005). The Declaration places utmost importance and attention on human rights, a concept that originated from Western societies. But there seems to be conflicting claims about human rights in Asia – one claim is that there is no room for human rights in Asian ethos, as Asians tend to value holistic happiness and welfare of the total group or community to which they belong rather than on their individual interests; while the opposite view is that Asian societies are often in favour, rather than against, human dignity and human rights

(Bagheri 2011). Despite existing controversy, Bagheri rightly pointed out that the Declaration was nevertheless adopted unanimously by member states.

Another limitation in the applicability of the UDBHR was raised by Macklin, who thought that Article 4 on the maximisation of direct and indirect benefits could possibly restrict current practice (Macklin 2005). Although this principle would be valid for clinical research that has some potential of direct or indirect benefits to research participants, it would seem to preclude biomedical research which is not intended to benefit research participants. Such research includes baseline physiological studies, examination of physiological mechanisms, and comparison studies in which invasive interventions are performed on both healthy research participants and patients. Even though there are no likely direct or indirect benefits for research participants, these are crucial studies that may give rise to generalizable knowledge and contribute to the common good. However, Macklin pointed out that the Declaration has managed this situation well by including Article 26 which states, "This Declaration is to be understood as a whole and the principles are to be understood as complementary and interrelated. Each principle is to be considered in the context of the other principles, as appropriate and relevant in the circumstances". The issue could be addressed by Article 5 of the Declaration, which states, "The autonomy of persons to make decisions, while taking responsibility for those decisions and respecting the autonomy of others, is to be respected". This means that participants who give voluntary, informed consent may participate in any biomedical research, provided a research ethics committee has assessed that the research protocol imposes acceptable risks in relation to the scientific value of the study, notwithstanding the absence of direct benefit for participants. Participants' autonomy should be respected if they agree to participate after being properly informed of the nature, risks and absence of direct benefit of the study. As such, Article 5 would moderate the benefit requirement of Article 4.

It has also been argued that solidarity, the principle articulated in Article 13 of the UDBHR, cannot be applied at a global level because it presupposes the existence of opposition groups. According to Heyd, "Since solidarity is created in the struggle for a collective cause, it is necessarily exclusive, presupposing the existence of competing causes…, solidarity is a social bonding that is formed against, or at least in competition with, other groups (Heyd 2007)." Thus, a global or universal solidarity is not possible since there is no universal value that people may identify with (Gunson 2009). Solidarity has also been said to be a peculiarly continental European value as it applies only to specific European practices, particularly in the European Welfare State, whereby everyone makes a fair contribution in return for equal access to healthcare as well as unemployment benefits, pensions and other goods (Houtepen and ter Meulen 2000). But is it impossible for all human beings to share a collective cause? As mentioned above, surely there must be common basic entitlements that all human beings desire, for instance, access to adequate nutrition and water (Article 14 -2-b). Although there may be competition between groups to gain access to such basic necessity, the cause itself is clearly a universal *human* aspiration. And solidarity, as commonly understood to refer to the ties that bind a society together, is not a foreign concept in non-Western societies. Its emphasis on the community, and a common good, is similar to Confucian thinking that pervades many Asian societies. As such, solidarity is not exclusive to the Europeans, but rather, a universal concept that is familiar to all citizens who share a sense of camaraderie with their fellow countrymen.

4.8 Further Elaboration on Principles Articulated in the UDBHR

The UDBHR was formulated with the aim of achieving consensus from all UNESCO member states. Although initially drafted by the IBC, which is comprised of bioethics experts, the contents were subject to substantial editing by governmental officials of the IGBC whose negotiation produced an instrument with "principles framed at a very high level of abstraction" (Snead 2009). As such, the UDBHR has been criticised for being "at best a toothless statement of vague principles, and at worst a potential source of mischief that will harm research and public health efforts" (Wolinsky 2006).

The IBC has since elaborated on the principles listed in the UDBHR, and issued several reports, including those on *Consent* (2008), *Social Responsibility and Health* (2010), and *Respect for Human Vulnerability and Personal Integrity* (2013). Most recently, the IBC has issued a draft report on the Principle on *Non-discrimination and Non-stigmatization* (2013). By expounding on the principles and their applications in detail, these reports offer a more comprehensive understanding of the UDBHR. For instance, although the UDBHR requires that special protection be accorded to persons without the capacity to consent (Article 7), it does not consider how lack of capacity could be determined, nor clarify who such persons may be. The 2008 *Consent Report* addresses this issue, and more. In the *Social Responsibility and Health Report*, the IBC acknowledges that Article 14 of the UDBHR is complex, and in attempt to further understanding and application, it provides elaboration on the ethical and legal dimensions of the principle, as well as illustrates how the principle could be translated into action through concrete examples. Similarly, the *Human Vulnerability and Personal Integrity Report* aims to promote the dissemination of the UDBHR, by enhancing the debate on Article 8 of the UDBHR. It addresses the problem of vagueness in the UDBHR, by providing greater clarity on the concept of "human vulnerability".

UNESCO also published a book in 2009 entitled *The UNESCO Universal Declaration on Bioethics and Human Rights: Background, Principles and Applications*. It provides a thorough examination of the Declaration, including an article-by-article explanation of how the instrument could be used as a tool to address ethical issues. As almost all of the authors were involved in the elaboration of the UDBHR, their contributions reveal the historical background of the instrument, the intention behind the inclusion of each article, as well as the potential interpretation and application of the Declaration.

Prior to 2005, UNESCO had produced other relevant documents, which may provide further insights into the principles mentioned in UDBHR. The *Declaration on the Responsibilities of the Present Generations Towards Future Generations* (1997) reminds the present generation that the needs and interests of future generations must be safeguarded. This Declaration therefore offers an elaboration of the principles articulated as Articles 16 and 17 of the UDBHR on protection of future generations and the environment. Before the UDBHR, the IBC was tasked with the preparatory work for the *Universal Declaration on the Human Genome and Human Rights* (1997), and the *International Declaration on Human Genetic Data* (2003). The *Human Genome Declaration* was the first international instrument to establish a legal and ethics framework for research on the human genome and the applications of findings (Lenoir 1998–1999), while the *Human Genetic Data Declaration* provides international guidelines for the collection, processing, use and storage of human genetic data, human proteomic data and biological samples. The need to respect human dignity and human rights was emphasised in both Declarations, although with a focus exclusively on the field of human genetics research. They therefore lack the broad application of the UDBHR, but nevertheless illustrate the possible application of international human rights law in a specific area in medical and scientific research. The UBDHR should be read together with these supplementary advices, for a more complete understanding of the intention and application of the espoused principles.

Finally, the United Nation's 1948 *Universal Declaration of Human Rights* is another key supplementary reading that sheds light on the concept of respect for human dignity and human rights, which is foundational to the UDBHR. In fact, the *Human Rights Declaration* is generally considered to be the foundation of international human rights law as it "has inspired more than 80 international human rights treaties and declarations, a great number of regional human rights conventions, domestic human rights bills and constitutional provisions, which together constitute a comprehensive legally binding system for the promotion and protection of human rights" (United Nations).

4.9 Conclusion

There are certain features of the UDBHR that contribute to its universality. Respect for human dignity, human rights and fundamental freedoms had been extensively incorporated into the Declaration, and it is the first time that international human rights law was comprehensively extended into the field of bioethics. The UDBHR shifted the main focus of bioethics from respect for individual autonomy, to larger society, and even more broadly to humanity and beyond, as it also urges consideration for our future generations, and other living beings that we share our environment with. By focusing on core values or principles that all countries share, the UDBHR established guidelines that are useful to both developing and developed countries, especially for those involved in transnational research. Calling on the principle of solidarity, the Declaration encourages benefits sharing amongst all sectors of society, and even within the international community. It also promotes international cooperation, to achieve universal access to basic rights that all human beings ought to have. The principles espoused in the UDBHR have been shown to, and are expected to, withstand the test of time. As basic terms and principles were not explicitly defined in the Declaration, it allows for various interpretations by different countries, with the practicalities of implementation left to the nation states. The UDBHR further promotes respect for cultural diversity and pluralism, and provides for exceptions, so long as international human rights law is observed.

UNESCO has successfully elaborated an international instrument that identifies universal norms of bioethics, to ensure that there is a minimum ethical standard in the conduct of biomedical research and clinical practice worldwide. The Declaration is a major accomplishment by UNESCO, as it managed to strike a delicate balance between respect for cultural diversity and demand for universal consensus (Gunson 2009, p. 242), to achieve unanimous acceptance by all the member states in such a sensitive and divergent area as bioethics.

References

Andorno, R. 2007. Global bioethics at UNESCO: In defence of the Universal Declaration on Bioethics and Human Rights. *Journal of Medical Ethics* 33: 150–154.

Andorno, R. 2009. Human dignity and human rights as a common ground for a global bioethics. *Journal of Medicine and Philosophy* 34: 223–240.

Bagheri, A. 2011. The impact of the UNESCO declaration in Asian and global bioethics. *Asian Bioethics Review* 3(2): 52–64.

Council for International Organizations of Medical Sciences and World Health Organization. 2002. *International ethical guidelines for biomedical research involving human subjects.*

Council of Europe. 1997. *Convention for the protection of human rights and dignity of the human being with regard to the application of biology and medicine: Convention on human rights and biomedicine.*

Gunson, D. 2009. Solidarity and the Universal Declaration on Bioethics and Human Rights. *Journal of Medicine and Philosophy* 34: 241–260.

Hayry, M., and T. Takala. 2005. Human dignity, bioethics, and human rights. *Developing World Bioethics* 5(3): 225–233.

Heyd, D. 2007. Justice and solidarity: The contractarian case against global justice. *Journal of Social Philosophy* 38(1): 112–130.

Houtepen, R., and R. ter Meulen. 2000. New types of solidarity in the European welfare state. *Health Care Analysis* 8(4): 329–340.

Kirby, M. 2008–2009. Human rights and bioethics: The Universal Declaration of Human Rights and UNESCO Universal Declaration of Bioethics and Human Rights. *Journal of Contemporary Health Law and Policy* 25: 309–331.

Lenoir, N. 1998–1999. Universal declaration on the human genome and human rights: The first legal and ethical framework at the global level. *Columbia Human Rights Law Review* 30: 537–587.

Macklin, R. 2005. Yet another guideline? The UNESCO draft declaration. *Developing World Bioethics* 5(3): 244–250.

National Commission for the Protection of Human Subjects of Biomedical and Behavioral Research. 1979a. *Belmont report.*
National Commission for the Protection of Human Subjects of Biomedical and Behavioral Research. 1979b. *Belmont report: Ethical principles and guidelines for the protection of human subjects of research.* U.S. Department of Health and Human Services.
Nuremberg Code. 1949. *Trials of war criminals before the Nuremberg Military Tribunals under Control Council law no. 10*, 2: 181–182.
Snead, O.C. 2009. Bioethics and self-governance: The lessons of the Universal Declaration on Bioethics and Human Rights. *Journal of Medicine and Philosophy* 34: 204–222.
ten Have, H. 2006. The activities of UNESCO in the area of ethics. *Kennedy Institute of Ethics Journal* 16(4): 333–351.
ten Have, H.A., and M.S. Jean (Eds.). 2009. *The UNESCO Universal Declaration on Bioethics and Human Rights.* UNESCO Publishing.
UNESCO. 2003. Records of the General Conference, 32nd Session, Paris, 29 September to 17 October 2003, *v. 1: Resolutions.*
UNESCO. 2005. *Universal Declaration on Bioethics and Human Rights.*
UNESCO International Bioethics Committee. 2008. *Report of the International Bioethics Committee of UNESCO on consent.*
UNESCO International Bioethics Committee. 2010. *Report of the International Bioethics Committee of UNESCO on social responsibility and health.*
UNESCO International Bioethics Committee. 2013. *The principle of respect for human vulnerability and personal integrity.*
United Nations. 1948. *Universal Declaration of Human Rights.*
United Nations. (n.d.). *Human rights law.* Retrieved from The Universal Declaration of Human Rights: http://www.un.org/en/documents/udhr/hr_law.shtml
United Nations Commission on Human Rights. 2003. *Resolution 2003/69: Human rights and bioethics.*
Wolinsky, H. 2006. Bioethics for the world. *EMBO Reports* 7(4): 354–358.
World Medical Association. 1964, Revised 2013. *Declaration of Helsinki: Ethical principles for medical research involving human subjects.*

Chapter 5
Consent and Bioethics

Sheila A.M. McLean

Abstract The *Universal Declaration on Bioethics and Human Rights* contains several articles that relate specifically to the question of consent. Importantly, both the Declaration and the subsequent International Bioethics Committee *Report on Consent*, which is designed to explicate the relevant articles, link what was historically a legal concept to broader and more nuanced concepts such as respect for dignity, integrity and autonomy. The link between these concepts (and particularly that of autonomy) is explored in this contribution, as is the influence of the concept of consent on bioethics as a whole and the work of the International Bioethics Committee in particular.

5.1 Introduction

It is probably no exaggeration to say that consent has become one of the most significant concepts in bioethical discourse. Sourced originally from the law (Rothman 2001) prohibiting unwanted or unauthorised touching of one person by another – the law of battery, assault or trespass to the person – the concept's obvious attraction for the developing field of bioethics was its commitment to the inviolability of the human being. This view of the individual as worthy of both protection and respect was already present, albeit in a relatively elementary form, in some legal systems, but was forcefully dragged into the broader public consciousness by the revelations that followed the trials of Nazi doctors in the aftermath of the Second World War. The Nuremberg Code (1947) that resulted from these trials, and which was specifically addressed to the question of human experimentation, was unequivocal in its declaration that "The voluntary consent of the human subject is absolutely essential" (article 1). Although focused on human subject experimentation, the Code has had wider ramifications and influence, such that "[i]nformed consent has been an axiom of post-World War II clinical research and practice…" (Weindling 2011).

S.A.M. McLean (✉)
Emeritus Professor, Law and Ethics in Medicine, Glasgow University, Glasgow, UK
e-mail: Sheila.McLean@glasgow.ac.uk

The Nuremberg Code was followed by the United Nations Declaration on Human Rights (1948) which starts with the commitment that, "[a]ll human beings are born free and equal in dignity and rights. They are endowed with reason and conscience and should act towards one another in a spirit of brotherhood". Article 3 continues that "[e]veryone has the right to life, liberty and security of person". Subsequently, organisations such as the World Medical Association have promulgated specific ethical codes concerning human experimentation, while nations in Europe, Africa and elsewhere have signed up to regional and international human rights declarations, all of which incorporate commitment to the inviolability of the human being.

In bioethical literature, the basic legal concept of consent has been further elaborated and developed using concepts such as respect for integrity, respect for human dignity and respect for autonomy. While 'dignity' remains a difficult concept to describe with any clarity, its implications are obvious; it is about respecting others, their values, their person and their beliefs. Equally, integrity might seem a somewhat vague notion, but we can broadly understand its purpose and content. While by no means a simple concept, as will be seen later, autonomy (or respect for autonomy) arguably played a central role in the developing jurisprudence on consent and significantly influenced the developing discipline of bioethics. This has influenced not just human experimentation (or more accurately research), but it has also affected clinical practice, such that the need to respect patient autonomy by obtaining the consent of individual research participants and patients is now almost taken for granted by scientists, physicians and their professional associations. As UNESCO's International Bioethics Committee (IBC) stresses, "[c]onsent is one of the basic principles of bioethics because it is closely linked to the principle of autonomy and because it reflects affirmation of human rights and human dignity which are the core values of democratic societies." (IBC Report on Consent 2008, para 141).

It is worth briefly noting that, although there was some early case law on consent, its recognition as a fundamental predictor of legally and ethically valid practice developed relatively slowly. Medicine has a long history of paternalism; the idea that 'doctor knows best', simply put. Pellegrino and Thomasma (1987) have said that paternalism was "….the dominant, and indeed the accepted, model of the clinical relationship for most of medicine's history". In a paternalistic relationship, patients' rights could be, and often were, essentially unrecognised, or at least were not the primary concern. On the contrary, the assumption was that the doctor, as possessor of the relevant skill and information, was able, even entitled, to decide what was in the patient's interests and act accordingly. However, in the aftermath of Nuremberg, and with growth in recognition of the importance of human rights, it became clear that the paternalistic model was, as O'Neill says "defective and failed to establish an adequate context for reasonable trust"; an essential component of the physician/patient relationship (O'Neill 2002, 18). Whatever value, if any, the paternalistic model of medicine may have had in the past, it is clear that it no longer enjoys widespread support. Rather, it is recognised that:

...many (if not most) patients want to be treated as an interested participant in healthcare decisions; this is hardly surprising. For them the importance of control over their lives is that it permits them to make decisions that reflect their own values and encapsulate their own interests (McLean 2010, 14).

In light of the increased focus on patients' rights, the developing recognition of the importance of patient autonomy and the impact of the burgeoning discipline of bioethics, it is scarcely surprising that from its inception UNESCO's International Bioethics Committee has engaged with consent, autonomy and related concepts on a regular basis.

The importance of consent, both in theory and in practice, is reinforced by the fact that it is referred to (sometimes very often) in all bar one of the reports produced by the International Bioethics Committee, even those that predate the promulgation of the IBC's ground-breaking Declaration on Bioethics and Human Rights (2005). Indeed – unsurprisingly, perhaps given the subject matter – only the report on Food, Plant Biotechnology and Ethics (1996) contains no specific mention of consent.

As suggested earlier, consent in the IBC's conception is intimately linked to other bioethical concepts, such as autonomy and dignity. For example, its most recent report on Traditional Medicine Systems and Their Ethical Implications, says, "[t]he principles of autonomy, informed consent and respect for human dignity are inseparable" (2013, para 4.2.1).

5.2 The Universal Declaration on Bioethics and Human Rights

One of the IBC's most important actions must be the development and promulgation of the Declaration on Bioethics and Human Rights in 2005. This Declaration follows broadly in the footsteps of earlier documents, such as the Council of Europe's *Convention for the Protection of Human Rights and Dignity of the Human Being with regard to the Application of Biology and Medicine: Convention on Human Rights and Biomedicine* (1997). However, the IBC's formulation of the rights inherent in humanity is a global, as opposed to a regional, recognition of the importance of human rights in the healthcare setting and, being relatively recent, it allows for encapsulation of the developments in, and enhanced capacities of, modern medicine and the challenges posed by advances in technology. In terms of consent, and while the various articles of the Declaration are inter-dependent, the most directly relevant articles are as follows:

Article 3 – Human dignity and human rights

1. Human dignity, human rights and fundamental freedoms are to be fully respected.
2. The interests and welfare of the individual should have priority over the sole interest of science or society.

This general commitment to respect for the individual is then followed by three articles that refer specifically to autonomy and/or consent:

Article 5 – Autonomy and individual responsibility
The autonomy of persons to make decisions, while taking responsibility for those decisions and respecting the autonomy of others, is to be respected. For persons who are not capable of exercising autonomy, special measures are to be taken to protect their rights and interests.

Article 6 – Consent

1. Any preventive, diagnostic and therapeutic medical intervention is only to be carried out with the prior, free and informed consent of the person concerned, based on adequate information. The consent should, where appropriate, be express and may be withdrawn by the person concerned at any time and for any reason without disadvantage or prejudice.
2. Scientific research should only be carried out with the prior, free, express and informed consent of the person concerned. The information should be adequate, provided in a comprehensible form and should include modalities for withdrawal of consent. Consent may be withdrawn by the person concerned at any time and for any reason without any disadvantage or prejudice. Exceptions to this principle should be made only in accordance with ethical and legal standards adopted by States, consistent with the principles and provisions set out in this Declaration, in particular in Article 27, and international human rights law.
3. In appropriate cases of research carried out on a group of persons or a community, additional agreement of the legal representatives of the group or community concerned may be sought. In no case should a collective community agreement or the consent of a community leader or other authority substitute for an individual's informed consent.

Article 7 – Persons without the capacity to consent
In accordance with domestic law, special protection is to be given to persons who do not have the capacity to consent:

(a) authorization for research and medical practice should be obtained in accordance with the best interest of the person concerned and in accordance with domestic law. However, the person concerned should be involved to the greatest extent possible in the decision-making process of consent, as well as that of withdrawing consent;
(b) research should only be carried out for his or her direct health benefit, subject to the authorization and the protective conditions prescribed by law, and if there is no research alternative of comparable effectiveness with research participants able to consent. Research which does not have potential direct health benefit should only be undertaken by way of exception, with the utmost restraint, exposing the person only to a minimal risk and minimal burden and, if the research is expected to contribute to the health benefit of other persons in the same category, subject to the conditions prescribed by law and compatible with the protection of the individual's human rights. Refusal of such persons to take part in research should be respected.

5.3 Consent and the International Bioethics Committee

It was, of course, never intended that the Declaration should be a static or sterile document. Following its adoption, the IBC committed itself to future work exploring and explicating the articles of the Declaration in more depth. Given its centrality to respect for human dignity and personal autonomy, it is perhaps unsurprising that the first articles selected for consideration were those relating to consent. Importantly, the report on Consent (2008) based its conclusions firmly in the traditions of human rights law, as well as emphasising the need for vigilance in ensuring that human rights norms are truly respected:

> Informed consent is a fundamental principle that has marked the emergence of modern medical ethics based on personal autonomy. The need for informed consent in biomedical research was emphasized by the Nuremberg trials that revealed inhuman experimentation on prisoners in concentration camps. Its importance in the context of scientific research was further strengthened by many examples of unethical human research that continued even in the post-World War II period. In the clinical context, the importance of informed consent has been recognized as a consequence of the rising patients' rights movement and emerging biomedical technologies that emphasized the necessity to decide about the complex healthcare choices to be made by the patient him/herself. The introduction of the practice of informed consent has also transformed the traditional paternalistic health-care professional-patient relationship (General Framework, para 5).

Situating bioethical consideration of the healthcare professional/patient relationship firmly within the domain of human rights is of great significance. As Knowles says, this "means moving towards a more expansive understanding of the relationships between human health, medicine and the environment, socioeconomic and civil and political rights, and public health initiatives and human rights" (Knowles 2001, 260).

In its report, the IBC further says that "[t]he principle of consent is closely related to the principle of autonomy" (Art. 5 of the Declaration) and the affirmation of human rights and respect for human dignity" (Art. 3 of the Declaration). The very structure of the text of the Declaration clearly reflects this close link. "(2008, General Framework, para 6). As we have seen, article 6.1 of the Declaration states categorically that "[a]ny preventive, diagnostic and therapeutic medical intervention is only to be carried out with the prior, free and informed consent of the person concerned, based on adequate information". This language is repeated verbatim in the IBC's report on Social Responsibility and Health (2010), and in one form or another appears in virtually every report produced by the Committee. As the report on Consent says, "[t]he doctrine of informed consent is one of the most well-known elements of medical ethics and bioethics today and is a pivotal principle that guides contemporary healthcare and research practices" (Introductory Remarks, 7).

The report also includes a helpful reminder that its scope is not limited to the articles in the Declaration that specifically refer to consent. Rather, "[w]hilst this report focuses on Articles 6 and 7 of the Declaration which address the issue of consent, these articles shall not be considered and interpreted separately from the other articles of the Declaration". As stated in Article 26, all principles 'are to be understood as complementary and interrelated' and 'considered in the context of the other principles, as appropriate and relevant in the circumstances' (Introduction, Para 4).

Clearly, the doctrine of consent has to cover a wide range of situations, from individual clinical decisions, through population based research, to involvement in scientific and medical research. For that reason, its role and content may vary considerably. Given this, it was a significant step by the IBC to attempt to explicate further the general commitment to consent outlined in the Declaration, and which was also addressed in earlier IBC reports which cover topics as disparate as genetics and the use of embryonic stem cells in research. The report on Consent has also informed subsequent IBC reports, such as those on Social Responsibility and Health (2010) and on Respect for Human Vulnerability and Personal Integrity (2011).

In its attempts to elucidate what a real or 'informed' consent encapsulates, the report on Consent combines a theoretical and principled understanding of the concept with a more practical attempt to put some flesh on the bones of how such a consent can (or perhaps, more accurately *should*) be sought. Article 6 of the Declaration states that "informed" consent is to be "based on adequate information". The report on Consent takes this further by specifying in more depth what this means in practice. Thus, it says that:

> With regard to the consent of the patient with a view to medical intervention, some important elements should be taken into account:
>
> - the diagnosis and the prognosis;
> - the nature and the process of the intervention;
> - the expected benefits of the intervention;
> - the possible undesirable side effects of the intervention;
> - possibilities, benefits and risks of alternative interventions.

Other elements that also need to be taken into account concern the experience and capabilities of the professionals involved in the medical intervention and their possible financial benefit in cases where there might be conflict of interest' (para 13)

Restating that the interests protected by the doctrine of consent are fundamental to all of humanity, the Report also reminds us that any limitation placed on it must be "by law, consistent with international human rights law, including laws in the interests of public safety, for the investigation, detection and prosecution of criminal offences, for the protection of public health or for the protection of the rights and freedoms of others (as stated in Article 27 of the Declaration)" (para 13). This statement serves as a valuable reminder that the issues surrounding patient/subject involvement in healthcare or research are bigger than merely medical or scientific considerations. Rather, they are intimately linked to wider and arguably more fundamental questions such as those of respect for dignity, integrity and autonomy which have already been said to have significantly influenced bioethics in its treatment of consent. Equally, it reinforces the important role played by the law in protecting these values and the individuals to whom they apply. Even despite this effort to explain the essential components of consent more clearly, the report recognises that it is by no means a simple idea, particularly as it is a process based in more nuanced and complex principles such as autonomy. In a very real sense, then, the practical and personal implications of a commitment to the importance of consent will hinge – at least in part – on what we mean by autonomy. Indeed, this will also have a direct effect on what information is 'adequate' in evaluating a purported consent. As the IBC report records, "although informed consent has been widely accepted in ethical discourse, its meaning has nevertheless remained beyond clear definition, stimulating an intense debate on this subject at both international and national levels" (Introductory Remarks, p. 7). In fact, arguably, there are two primary debates. The first of these can be described as legal; the second as bioethical.

The legal debate is in many ways the easier to describe, even although it is by no means easy to resolve. Put simply, it is a matter of process. How jurisdictions develop legal tests to evaluate the validity of a purported consent will have a direct and generally definitive impact on its effect. Thus, it is important to recognise that the patient/physician relationship is not 'symmetrical' (report on Consent, para 17). It is, therefore, the responsibility of the person with the expertise and knowledge – the physician or researcher– to ensure that adequate information is provided in a manner that allows for the patient or research subject to understand it. As the report continues, the process of seeking consent must be more than merely an "administrative procedure or a legal obligation, but rather an acknowledgement of the trust placed in him/her by the patient" (para 18). While this is manifestly true, how the law judges the quality of any consent apparently obtained will be important in evaluating whether or not this 'acknowledgement of trust' has actually been achieved.

In some jurisdictions –such as the United Kingdom– the jurisprudence that has developed in this area is based on what is referred to as a 'prudent doctor' test. (*Hunter v Hanley* (1955); *Bolam v Friern Hospital Management Committee* 1957; *Sidaway v Board of Governors of the Bethlem Royal Hospital* (1985)) even although there are some signs that this may be changing (*Bolitho v City and Hackney Health Authority* (1997); *Pearce v United Bristol Healthcare NHS Trust* (1999); *Chester v Afshar* (2004)) courts are still often reliant on an evaluation of what is 'reasonable' medical practice; a test which depends to a considerable extent on examining what a 'prudent' doctor would have disclosed. This reliance inevitably has an impact on what is *in fact* disclosed, or more accurately on what courts think *should* be disclosed before a purported consent can be said to be valid in law. It is obvious that where a test such as this is used, the focus is not on the individual patient and his/her beliefs, circumstances and intentions but rather on what physicians believe it is necessary to disclose. While the examples provided by the IBC, and referred to above, are helpful in highlighting the kind of information that should be passed to a patient, how these are interpreted in law will affect their content and how carefully they are adhered to. Perhaps more importantly, the legal test used will have a clear and unequivocal impact on whether or not it is possible in reality to see the law of consent as a device which protects patient autonomy. While the IBC report, as we have seen, links autonomy and consent closely together, in practice, the approach of the law can create a dissonance between them.

Even in those jurisdictions, such as Australia (*Rogers v Whittaker* 1992), Canada (*Reibl v Hughes* 1980) and some US States (*Canterbury v Spence* 1972), which adhere to a 'prudent patient' rather than a 'prudent doctor' test, the terminology itself makes it clear that it is not the *instant* patient, but rather a kind of generalised patient, who is the model. Thus, while at first sight it might appear that this is a preferable test in terms of vindicating patients' rights, Skegg argues that even if this test is used "the interests of the overwhelming majority of patients would be totally unaffected" (Skegg 1999, 146). In Canada, Robertson reports that even after the landmark case of *Reibl v Hughes*, which established the use of a 'prudent patient' rather than a 'prudent doctor' test, "plaintiffs lose in the vast majority of informed consent cases" (Robertson 2003, 154). It would seem, then, that the law's practice of creating, perhaps even its need to develop, tests that can be uniformly applied across all cases where patients are aggrieved at a perceived failure to provide sufficient information to enable them to make an autonomous decision, stands resolutely in the way of a system that genuinely does protect individual autonomy. As has been said, "…the need to elaborate a normative formula that will ensure consistency of application and the certainty that people can legitimately expect of the law, may make it difficult to accommodate the subtleties and nuances of the ethical debate, or to inquire into the autonomy quotient of the decision" (McLean 2010, 66).

Moreover, the law's approach to autonomy is largely based on questions of competence or legal capacity. In other words, the assumption is that if a person is legally competent, then they are free to make their own 'autonomous' decisions following the adequate provision of information about such matters as the risks, benefits and alternatives in the proposed treatment. However, competent decisions and autonomous ones may not be the same. The IBC report on Consent says: "[a]utonomy is often described as self-rule and refers to the right of persons to make authentic choices about what they shall do, what shall be done to them and, as far as possible, what should happen to them" (para 77). The most important word in this quotation is 'authentic'; the critical issue is how authenticity is to be measured. The mere making of a decision does not necessarily fulfil the aspiration to achieve self-rule or authenticity. O'Neill, for example, argues that "[b]y insisting on the importance of informed consent we *make it possible* for individuals to choose autonomously but we in no way guarantee require that they do so" (O'Neill 2002, 2). While it is self-evident that "[i]n order to deliberate rationally, a patient must be able to comprehend the relevant facts and circumstances of her situation…" – a test very like that of competence – what is vital is that this allows the patient "to discern which option best coheres with her life plan" (Beste 2005, 200–201) – an outcome more akin to autonomy. Space constraints prohibit further analysis of this point, but it is an important consideration when attempting to translate the value of autonomy into a workable legal framework. As O'Neill says, "[w]hat passes for patient autonomy in medical practice is operationalized by practices of informed consent" (O'Neill 2002, 38). However, if the processes and tests adopted by law rely heavily on assessments of competence rather than autonomy, then they will be ill-suited to scrutinising decisions for their authenticity.

The bioethical debate is – if anything – even more complex. While it has been suggested that the law may struggle in practice to reflect or address the autonomy of the individual, bioethical literature highlights that there is no agreement on precisely what autonomy means. Some theories of autonomy are firmly rooted in the individual – as would seem to be the approach adopted by the Universal Declaration which, in common with other similar international statements, refers to the dignity, integrity and autonomy *of the individual* as being of fundamental importance. This approach emphasises that it is the right of every individual to make decisions that are personally coherent; that reflect their own values. As Harris says, "….autonomy itself is part of our concept of the person because it is autonomy that enables the individual to make her life her own" (Harris 2003, 10).

Still other theories, such as that of John Stuart Mill, also emphasise the right of the individual to act free from constraints imposed by others, albeit that the freedom to act is constrained by the need to ensure that it does not cause harm to third parties. As he argues:

> ….the sole end for which mankind are warranted, individually or collectively, in interfering with the liberty of action of any of their number, is self-protection….the only purpose for which power can be rightfully exercised over any member of a civilized community, against his will, is to prevent harm to others. (Mill 1869, transcript 9)

For Kant (1785), "autonomy is manifested in a life in which duties are met, in which there is respect for others and their rights, rather than in a life liberated from all bonds…it is a matter of acting on certain sorts of principles, and specifically on principles of obligation" (O'Neill 2002, 83–84). Autonomy, in Kant's ideology, is rooted in duties rather than rights. That is, to be autonomous is "emphatically not to be able to do or have whatever one desires, but rather it is to have the capacity for rational self-governance" (Downie and Macnaughton 2007, 42).

For some feminist theorists, and others, autonomy is best seen as a relational rather than an individual concept. This has two implications. First, it is argued that as we are "embedded in social relations someone else who shares my culture might be able to understand me better than I understand myself" (Williams 2005). This would suggest that the mere capacity to make a decision should not be seen as equivalent to autonomy; rather we share assumptions and commitments that shape these decisions. The second facet of the relational account is its concern for the impact of our decisions on others. Rejecting what has been called 'moral atomism', (Pellegrino and Thomasma 1987, 34). it is argued that "[a]ny conception of autonomy that fails to incorporate socially situated interpersonal relations rests on illusion" (Donchin 2000, 189).

While this is an extremely brief account of various approaches to autonomy and what constitutes an autonomous decision, and it is by no means exhaustive of them, it is nonetheless important to recognise that there is no widespread agreement on what these concepts mean. For the IBC, what might be seen as something of a hybrid definition is used. The Declaration takes an approach very similar to that advocated by J.S. Mill, while the report on Consent recasts this somewhat, saying "[a]utonomy implies responsibility. The power to decide for one's self entails *ipso facto* acceptance of the consequences of one's actions, which, in health matters, can be awesome. Therefore, it should be emphasized that the person needs to be informed of the precise consequences of his/her choice…" (Framework, para 7).

5.4 The Declaration, the Consent Report and Additional Issues

While the focus of this discussion has primarily been on the competent adult person, both the Declaration and the report on Consent consider wider issues. Space does not permit more than an acknowledgement of them here, but this is not to suggest that they lack importance. As has already been seen, article 7 of the Declaration directly considers those who are deemed to lack capacity to consent to treatment – who are not seen as autonomous. In the report on Consent, these individuals and groups are described as being "those who, for reasons internal to themselves, do not have the capacity to make autonomous choices irrespective of their external circumstances. Various groups of people have been traditionally labelled in this way. They include people with learning difficulties, the mentally ill, children, confused elderly and unconscious people" (para 78). For these groups, additional safeguards will be necessary to ensure that their interests are protected. However, the report also makes it very clear that "no judgment of capacity to consent should be called for unless there is evidence to undermine the normal assumption that people are able to decide for themselves. In other words, proof of incapacity is required not proof of capacity. Foolish decisions can be voluntarily made by the most autonomous people and the freedom to do so should not be restricted by imposing over-stringent standards of capacity" (para 80). This, the report continues, means that a simplistic approach is inappropriate as the criteria that apply to capacity are not necessarily objective. It is not, for example, tantamount to lack of capacity simply that someone else disagrees with the individual's decision, nor that it appears irrational to third parties. Risks and benefits can and often may be interpreted differently by different people. As an example, the report says:

> ...a patient might not wish to receive possible life-saving treatment for a malignant disease but rather maximize the quality of their remaining days by avoiding the rigours of cytotoxic medication. To interpret such an outcome as unreasonable would compromise the consent process for if the patient chooses the treatment he/she will be regarded as able to consent and so undergo the procedure and if he refuses then the procedure will still be carried out as the unreasonable choice will indicate his/her incapacity to consent and thus invalidate his/her refusal (para 81 (ii)).

The report also addresses how consent is demonstrated, and important questions surrounding the respect to be given to advance directives (statements) and donation of organs for transplantation.

5.5 Conclusion

The report on Consent is manifestly extremely wide-ranging, only parts of which have been considered in any depth in this discussion. However, there is a reason for having adopted this approach; most importantly, only by understanding the complexities associated with describing autonomy, and by interrogating what makes for an autonomous decision, is it possible to move to consideration of what it means not to be autonomous. The former are essential to understanding the latter. Equally, no discussion of the *form* in which consent is demonstrated can logically precede an analysis of what consent actually amounts to and encapsulates, even if it may be important in recording whether or not consent has been given (at least in theory).

The IBC's Universal Declaration on Bioethics and Human Rights seeks to establish a normative framework. The further work contained in the report on Consent is aimed at guiding states as to how these norms might be translated into their culture and laws. As this discussion has suggested, however, this latter effort is both ambitious and complex. Where no agreement exists on what precisely constitutes an adequate definition of autonomy, and with jurisprudence adopting different approaches to testing the quality of a purported consent, it is evident that guidelines – like the concepts themselves – are subject to interpretation. Indeed, the report on Consent explicitly recognises this. Whereas many of the so-called liberal, Western democracies – at least in theory – adopt a relatively individualistic account of autonomy, other cultures approach this from a very different perspective – arguably a more relational one that may involve family or community elders in healthcare decisions. In such situations, the report recognises the "difficulty in aligning the autonomy of individuals that is embodied in Article 5…" (para 114). While reinforcing that, even in these situations "seeking consent from an individual is indispensable", the report nonetheless concludes that "the actual value of the consent of an individual, once the community has given its approval, may sometimes provoke questioning" (para 115). For the IBC, "[i]t is necessary that the issue of consent be envisaged in a more global context of education and making persons autonomous whilst keeping in mind the primacy of the interests of the person concerned in their social setting. It is necessary to ensure the respect for the will of the person concerned, and to promote education towards autonomy and individual responsibility" (para 120). Focusing on individuals rather than communities will not sit comfortably in every social or cultural setting. Nonetheless, even when this is the case, the individual remains important, and the norms of respect, integrity and autonomy do not shrivel into insignificance. Rather, they may become more nuanced, more culturally sensitive and more socially appropriate.

Despite the potential difficulties highlighted by this discussion, which are after all common to all international guidelines, the IBC's attempt to elaborate, from a principled perspective, the subject of consent in healthcare and human subject research performs a valuable function in focusing those to whom it is addressed on its fundamental importance and providing guidance as to how the values it incorporates might be translated into reality. As the report says:

> The interpretation and implementation of the principle of consent as stated in Articles 6 and 7 of the Declaration definitely require the active participation of States. These articles should serve as a framework for legislation, regulations and policy decisions within the Member States. Moreover, since experience in many domains has shown that laws or regulations are only effectively enforced if they are backed by action in education, training and information, States should also have a specific responsibility in promoting education, training and information in the fields relevant to bioethics (para 137).

It is in clarification of complex concepts and provision of guidance as to the implementation of the aspirations that accompany them that the IBC's interpretation of the articles of the Declaration on Bioethics and Human Rights has a vital role to play. However autonomy is conceptualised, and no matter how its legal parallel of consent is approached in law, the IBC's reaffirmation of its significance can and should inform and guide states, legislatures and individuals in their efforts to meet and give effect to the requirements of international human rights norms and law. There can be few, if any, more important goals.

References

Beste, J. 2005. Instilling hope and respecting patient autonomy: Reconciling apparently conflicting duties. *Bioethics* 19(3): 215–251.
Donchin, A. 2000. Autonomy, interdependence, and assisted suicide: Respecting boundaries/crossing lines. *Bioethics* 14(3): 187–204.
Downie, R.S., and J. Macnaughton. 2007. *Bioethics and the humanities: Attitudes and perceptions.* Abingdon: Routledge-Cavendish.
Harris, J. 2003. Consent and end of life decisions. *Journal of Medical Ethics* 29: 10–15.
Kant, I. 1785. *Fundamental principles of the metaphysics of morals.*
Knowles, L.P. 2001. The Lingua Franca of human rights and the rise of a global bioethics. *Cambridge Quarterly of Healthcare Ethics* 10: 253–263.
McLean, S.A.M. 2010. *Autonomy, consent and the law*. London: Routledge.
Mill, J.S. 1869. *On liberty*, transcript at p. 9. Available at: http://www.bartleby.com/130/1.html. Accessed 7 Oct 2013.
O'Neill, O. 2002. *Autonomy and trust in bioethics*. Cambridge: Cambridge University Press.
Pellegrino, E.D., and D.C. Thomasma. 1987. The conflict between autonomy and beneficence in medical ethics: Proposal for a resolution. *Journal of Contemporary Health Law and Policy* 3(23): 23–46.
Robertson, G. 2003. Informed consent: 20 years later. *Health Law Journal Special Edition*, 153–159.
Rothman, D.J. 2001. The origins and consequences of patient autonomy: A 25-year retrospective. *Health Care Analysis* 9: 255–263.
Skegg, P.D.G. 1999. English medical law and "informed consent": An antipodean assessment and alternative. *Medical Law Review* 7(Summer): 135–165.
UNESCO IBC report on Consent 2008. Available at: http://unesdoc.unesco.org/images/0017/001781/178124e.pdf. Last visited 2 May 2014.
UNESCO Universal Declaration on Bioethics and Human Rights. 2005. Available at: http://www.unesco.org/new/en/social-and-human-sciences/themes/bioethics/bioethics-and-human-rights/. Last visited 2 May 2014.
Weindling, P. 2011. The origins of informed consent: The International Scientific Commission on War Crimes, and the Nuremberg Code. *Bulletin of the History of Medicine* 75: 37–71 (at p. 37).
Williams, S.H. 2005. Comment: Autonomy and the public-private distinction in bioethics and law. Available at: http://muse.jhu.edu/journals/indiana_journal_of_global_legal_studies/v012/12.2williams_s.pdf. Accessed 7 Oct 2013.

Table of Cases

Bolam v Friern Hospital Management Committee [1957] 2 All ER 118
Bolitho v City and Hackney Health Authority [1997] 4 All ER 771
Canterbury v Spence 464 F 2d 772 (DC 1972)
Chester v Afshar [2004] 4 All ER 587
Hunter v Hanley 1955 SC 200

Pearce v United Bristol Healthcare NHS Trust (1999) 48 BMLR 118
Reibl v Hughes [1980] 2 SCR 880
Rogers v Whittaker (1994) 16 BMLR 148 (1992)
Sidaway v Board of Governors of the Bethlem Royal Hospital [1985] 1 All ER 643

Chapter 6
Global Bioethics as *Social* Bioethics

Stefano Semplici

Abstract According to Article 1 of the *Universal Declaration on Bioethics and Human Rights* of 2005, bioethics "addresses ethical issues related to medicine, life sciences and associated technologies as applied to human beings, taking into account their social, legal and environmental dimensions". This definition broadens the scope of the discipline, far beyond the content of the traditional and more controversial issues concerning the beginning and the end of life or the limits of research. The right of every individual to enjoy the highest attainable standard of health is acknowledged – among others – as one of the principles that *global* bioethics must comply with, including all the determinants of human development and well-being. Therefore, social responsibility and respect for persons and groups living in conditions of special vulnerability, knowledge and benefit sharing, and sustainable development are key in the work of the UNESCO International Bioethics Committee (IBC). The commitment to improving the standard of health, dignity and quality of life for every human being is a matter of society as well as a matter of science.

6.1 Introduction

The definition of bioethics proposed by W.T. Reich in the *Introduction* to the second edition of the *Encyclopedia of Bioethics* (1995) seems to grasp both the complexity and the breadth of the new discipline: bioethics is "the systematic study of the moral dimensions – including moral vision, decisions, conduct, and policies – of the life sciences and health care". This study – the definition continues to explain – is further characterized by "employing a variety of ethical methodologies in an interdisciplinary setting". We have therefore a double scope to explore (life sciences and health care) – together with a commitment to focus on the moral dimensions implied therein – and the methodological direction of an interdisciplinary approach. If we compare this very standard definition with the one provided in Article 1 of the

---

S. Semplici (✉)
Social Ethics, University of Rome "Tor Vergata", Rome, Italy
e-mail: semplici@lettere.uniroma2.it

Universal Declaration on Bioethics and Human Rights of 2005, it is easy to point out a relevant difference. Bioethics, so the text reads, "addresses ethical issues related to medicine, life sciences and associated technologies as applied to human beings, taking into account their social, legal and environmental dimensions". *As applied to human beings.* This sounds quite obvious, in a document whose aim is exactly to link bioethics and *human* rights. The double scope is the same, although the order of the terms, not coincidentally, is reversed. The idea that bioethics addresses moral issues remains unchallenged: bioethicists are still called on to consider "human conduct", to quote the definition already proposed in 1978 in the first edition of the *Encyclopedia*. At the same time, however, the *Declaration* focuses on the consequences of scientific developments and their technological applications *on human beings*. Even though Article 17 calls for protection of the environment, the biosphere and biodiversity, building on the awareness of our interconnection with "other forms of life", it is exactly the destiny of human beings which appears to be the pivotal content of the ethical responsibility which is the cornerstone of *this* bioethics.

This is why the UNESCO definition can be and has indeed been contended as a narrow one. It is also true, however, that it broadens the scope of bioethics in a decisive way, by deepening the reference to health and health care as well as to the several determinants of the context in which the *application* of science and new technologies takes place. The right of every individual to enjoy the highest attainable standard of health, already acknowledged in the *Constitution* itself of the World Health Organization, is included among the principles with which *universal* bioethics must comply. As a consequence, access to quality health care – together with access to adequate nutrition and water, improvement of living conditions, elimination of the marginalization and exclusion of persons on any basis, and reduction of poverty and illiteracy – ought necessarily to be considered as one of the main ethical and political challenges entailed in the progress of science and technology. The task goes far beyond the content of the traditional and more controversial issues concerning the beginning and the end of life or the limits of research. Inasmuch as bioethics is about health and health care, it is at the very crossroads of all the determinants of human development and well-being. This *social* dimension of bioethics is explicitly highlighted in Article 14 of the *Declaration*. The very history of this article is evidence of its novelty and importance (Martinez Palomo 2009). It was not included in the first drafts of the document and was added to point out the necessity to go beyond the limits of purely medical ethics in order to "place bioethics and scientific progress within the context of reflection open to the political and social world" (UNESCO 2010, p. 9). It is all the more significant to observe that this principle of social responsibility and health was the second one, after the principle of consent, to be further investigated by the International Bioethics Committee, which published a Report on the topic.

The acknowledgment of everyone's right to enjoy the highest attainable standard of health has been long since an unquestionable premise for many declarations and other texts of international importance, as I have just underscored. The same applies to the interconnectedness of health with the other indispensable determinants of human development. Article 25 of the *Declaration* of 1948, a building block for – among others – Article 12 of the *International Covenant on Economic, Social and Cultural Rights* of 1966, had already made clear not only that everyone has the right to a "standard of living adequate for the health and well-being of himself and of his family", but also that, precisely in order to flesh out this fundamental right, it is necessary to provide "food, clothing, housing and medical care and necessary social services". In the first chapter of the *Human Development Report 1990*, it was affirmed that the choice of life expectancy as one of the indexes was motivated exactly by its functioning as a "proxy measure for several other important variables", such as "adequate nutrition, good health and education" (UNDP 1990, p. 11), whose overlap was thus implicitly reaffirmed. And many other examples may follow. In the *Report* titled *Closing the gap in a generation* and published in 2008 by the Commission established by WHO in 2005, the same year of the adoption of the UNESCO Declaration, a "holistic view" was strongly recommended, based on the awareness that, "the poor health of the poor, the social gradient in health within countries, and the marked health inequities between countries are caused by the unequal distribution of power, income, goods, and services, globally and nationally, the consequent unfairness in the immediate, visible circumstances of people's lives – their access to health care, schools, and education, their conditions of work and leisure, their homes, communities, towns, or cities – and their chances of leading a flourishing life" (WHO 2008, p. 1). The *Better Life Initiative*, launched in 2011 by the Organization for Economic Cooperation and Development with the aim of helping governments design better policies for better lives for their citizens, points out 11 dimensions which are identified as essentials to well-being. It goes without saying that health is one of them, together with housing, income, jobs, community, education, environment, civic engagement, life satisfaction, safety, and work-life balance (http://www.oecdbetterlifeindex.org/).

6.2 Social Responsibility and Social Vulnerability

The novelty in Article 14 of the *Declaration* of 2005 is exactly the use of the concept of *social* responsibility: "The promotion of health and social development for their people is a central purpose of governments that all sectors of society share". *All sectors of society*. The action of governments remains the cornerstone of the comprehensive strategy required to address the blatant injustice that so many people are still prevented from enjoying not the highest, but even a reasonable standard of health. What is affirmed is that an efficient, avoidable and affordable system of health care would be insufficient to perform the task, if the social determinants of health are not also targeted. In addition to this more obvious observation, it is underscored that social actors must play a role as important as the role of governments: individuals, groups and associations with different origins and missions, media, enterprises.

This observation becomes a key for the *social* bioethics indicated by UNESCO and especially by the IBC in a double sense. An illustrative example of the first one is provided exactly by the reference to the economic and industrial activity. As it is well known, this is the field in which the notion of social responsibility was first introduced, with the aim of reshaping the mission of management and taking into account the broad pervasiveness of the effects of its decisions as well as of the processes of production and distribution of goods and services. According to this new approach, a shift is required from the responsibility of just maximizing profit for the *stock* holders to the responsibility of respecting the rights and interests of all other *stake* holders affected by the enterprise's activity: customers, suppliers, employees, but also people living close to a plant, environmentalists, and other special interest groups. This is the most debated and the most successful aspect of the so-called corporate social responsibility: there are many negative 'externalities' produced not only by 'heavy' industry, and the actors of today's global market ought to include them in their plans and strategies.

Industry is one of the "special areas of focus" considered in the *Report* of the IBC on social responsibility and health and the potential 'dark side' of its activity is clearly indicated: "Work conditions can be harmful for people. Pollution can damage the environment and jeopardize the well-being of the population. Marketing strategies are often used to boost unhealthy behavior related to food and lifestyles. Research itself may serve profit-oriented activities more than interests and needs of individuals and society, exposing experts to conflict of interests that are always very dangerous because of their influence in decision-making processes. Globalization has made these risks more evident and has made the traditional institutional means of control less effective" (UNESCO 2010, pp. 31–32). The next step, considering the many social determinants and therefore the many social actors whose choices and actions rebound on people's health, is to necessarily include all of them in this call to share old and new responsibilities. It is sufficient, once again, to think of the role of education, which provides knowledge and awareness to better manage one's own life and to not run avoidable risks, as well as to claim the rights that are otherwise likely to remain just an object of wishful thinking, if not lip service. It is not only about educators and scientists. The last word of the *Report* is for the media, which are "in a position to be very helpful in sensitizing the population to health challenges and in explaining widely current questions and their societal dimensions", provided that they avoid their own temptations, such as seeking notoriety by triggering sensationalism, alarmism, or "even panic" (UNESCO 2010, p. 42).

The principle of externalities is one of the six principles enumerated by R.E. Freeman, the father of the stakeholder approach, as the ground rules of his doctrine of fair contracts. In this perspective, it is defined as follows: "If a contract between A and B imposes a cost on C, then C has the option to become a party to the contract, and the terms are renegotiated". The principle of governance is another one of these ground rules: "The procedure for changing the rules of the game must be agreed upon by unanimous consent". Other principles (entry and exit, contracting costs, agency) are also conceived as elements of a contractual theory, including the last one, which is the principle of limited immortality: although stakeholders are necessarily uncertain about the future, "the corporation shall be managed as if it can continue to serve the interests of stakeholders through time" (Freeman 1994, p. 417). This clarification helps focus on the second, decisive point to make. Of course, social responsibility calls on everyone to take into due account the consequences of their actions. However, this is not to think as if the task to perform were simply that of grounding the fair conditions of symmetry in a 'contractual' relationship. We could be even less satisfied by tracing back the concept to the juridical framework of the imputability for the damage caused to others.

The responsibility we are talking about implies unmistakably some *positive* obligations: Article 14 calls on all sectors of society to *promote* health and social development and not just to refrain from doing something. It means that all sectors of society are urged to boost the *positive* externalities of their activity and consider them as a goal to include in their plans of action. It means, at the same time, that governance is key in terms of *empowerment* of every single human being. In the language of politics and institutions, such a sharing of responsibilities entails a turn towards what we could define, following R.A. Dahl's successful intuition on the outcome of modern democracy (Dahl 1971), a *polyarchic* approach to the duty to respect, protect and fulfil fundamental human rights. In the language of philosophy, it is unavoidable to refer to – among others – the concept of responsibility established by P. Ricoeur, who builds on Jonas' work to propose a concept of responsibility enshrined in the experience of human vulnerability: we are responsible, even before our deeds, for the vulnerability of others. Therefore we ought to take care of them (Ricoeur 2004, Chap. II.4).

Not coincidentally, the respect for human vulnerability and personal integrity is also a principle stated in the *Universal Declaration on Bioethics and Human Rights*, and it has been addressed in the Report finalized by the IBC in 2011 (immediately after the Report on social responsibility for health) and published in 2013. Humankind as such is vulnerable, and we all may happen to lack at some point the ability and/or the resources which are necessary to prevent 'wounds' to our physical or mental integrity. In order to protect everyone's health, governments and all sectors and actors of society are called on not only to reduce the risks engendered by this *anthropological* vulnerability, which is by itself a powerful driver of solidarity, but also to effectively address those conditions of *special* vulnerability which create many faults of inequality that obstruct the enjoyment of a fundamental human right. Two categories are highlighted. On the one hand, there are *individual* conditions linked to temporary or permanent disabilities, diseases and limitations imposed by the stages of human life. On the other hand, there are those "social, political and environmental determinants" which are likely to expose people to an increased *collective* risk of vulnerability: "Many individuals, groups and population nowadays become especially vulnerable because of factors created and implemented by other human beings [...] Social vulnerability plays a role not only in biomedical research but also in the healthcare setting and in the development, implementation and application of emerging technologies in biomedical sciences and is a fact of life for a considerable portion of world's population". Poverty and inequalities "in income, social conditions, education and access to information", are mentioned as the first examples of such determinants of special vulnerability. They have an impact even on the capacity to prevent or at least ameliorate the effects of natural disasters: living in a country free from the risk of heavy earthquakes, for instance, is by all evidence not the same as living in those countries that have been devastated by them throughout their history. At the same time, however, living in a country where the most advanced technologies to cope with this risk are available, as well as the resources which make them affordable and the political determination to turn this knowledge into practice, is not the same as living in a country which lacks some of these elements, if not all of them (UNESCO 2013, pp. 14–15). Knowledge, wealth, governance. Once again, their improvement and fair distribution come out to be the pillars of new bioethical responsibilities, inasmuch as they are the fundamental determinants of human development. They overlap each other. Suffice it to mention some of the illustrative examples that are proposed in this Report of the IBC and related to the three specific domains pointed out in Article 8 of the *Declaration*: the doctor-patient relationship in the clinical setting, the researcher-subject relationship in human subject research, and the development and application of emerging technologies in the biomedical sciences. Neglected tropical diseases are parasitic and bacterial diseases that affect some of the most impoverished populations of the world: pharmaceutical companies show little interest in their treatment because of the lack of return on investment. Poverty is always a condition which jeopardizes the access to health care on equal footing as well as the real freedom to volunteer for research. The lack of regulation offers too often the possibility for vested interests and practices of exploitation to step into the breach. This is why we cannot hope to comply with our old and new bioethical obligations by simply focusing on one single determinant. An all-encompassing approach to human development is required and the issue of health equity is no exception to the rule. The IBC reaffirms that "we cannot be satisfied with the simple exercise of restraint and forbearance in pursuing our own objectives when this might threaten the autonomy and dignity of others. We are compelled to act in a positive way to help other people to cope with the natural or social determinants of vulnerability […] There is no doubt that empowerment of people against vulnerability entails more resources available for everyone, free and safe living conditions, access to quality health care as condition to actually guarantee to every human being 'the enjoyment of the highest attainable standard of health'" (Art. 14 of the Declaration). In this sense, the respect for human vulnerability and personal integrity is at the crossroads of unavoidable political responsibilities (UNESCO 2013, p. 14).

6.3 Is Knowledge an Appropriable Good?

A global bioethics genuinely committed to promoting health and social development is necessarily entrusted with reversing the "toxic combination" of those "structural determinants and conditions of daily life" which "are responsible for a major part of health inequities between and within countries" (WHO 2008, p. 1), so that the same combination can be eventually transformed into a *fruitful* one. Needless to say, the capacity to improve both education and 'the wealth of nations' is key, but it does not encompass all the aspects we are talking about. The *social* framework on which UNESCO and the IBC have insisted over these last two decades is obviously in line with this general yet crucial observation. We can probably try and make a step forward to the future by focusing on three points whose relevance is also highlighted in the *Declaration* of 2005. They are connected to each other through the idea of *sharing*, which is obviously pivotal to flesh out the human rights approach and therefore the universalism as well as the concept of equal dignity that it entails. These three points for a possible agenda are: the principle of benefit sharing, whose most striking example remains the persistent trade-off between the protection of intellectual property and the right of every human being to receive quality health care according to his or her needs; the building and dissemination of the capacity to actively participate in the 'production' of knowledge and science; and the link between health and well-being of people and education to sustainable development.

In the *Declaration* of 2005, the article devoted to the principle of benefit sharing comes immediately after the article on social responsibility and health. The principle is very demanding because it requires more than a "special and sustainable assistance" to the persons and groups that have taken part in a research project. Such benefits ought to be shared "with society as a whole and within the international community, in particular with developing countries". Once again, the task is a global one, in terms – among others – of access to quality health care, provision of new diagnostic and therapeutic modalities or products stemming from research, support for health services, and access to scientific and technological knowledge. This principle addresses and calls attention to many demanding challenges, both at the domestic and the international level. It also underlies the fundamental question on whether and to what extent we are allowed to consider the knowledge on which life itself could depend as an appropriable good. This debate is likely to go on, exactly because of the unprecedented pace of scientific progress.

Suffice it to recall the example of the human genome. The decision of the US Supreme Court of June 13, 2013, on the patentability of human genes (Association for Molecular Pathology et al. v. Myriad Genetics, Inc., et al.) has set a real landmark in one of the most controversial ethical and juridical issues that have been discussed in recent years. The final and unanimous decision was that isolated human genes may not be patented. It appears to be a clear affirmation of the principle enshrined in *The Universal Declaration on the Human Genome and Human Rights* of 1997, according to which the human genome is to be considered "the heritage of humanity" (Art. 1). Such a definition seems to imply a strict duty: whatever possible invention relies in this case on something that cannot be appropriated, and Article 4 excludes the possibility of making profit from the human genome in an explicit and unquestionable way: "The human genome in its natural state shall not give raise to financial gains". In its *natural* state. This was the breach where patent claims immediately stepped in. And the breach remains open after this decision of the Court. It is essential to underline what is "not implicated" by this decision on the two genes whose mutations confer a high risk for breast and ovarian cancer. *Method* claims are not excluded: "Had Myriad created an innovative method of manipulating genes while searching for the BRCA1 and BRCA2 genes, it could possibly have sought a method patent". Secondly, "this case does not involve patents on new applications of knowledge about the BRCA1 and BRCA2 genes". Therefore, such claims remain in principle unchallenged. "Nor do we consider the patentability of DNA – so the text of the ruling goes on to specify – in which the order of the naturally occurring nucleotides has been altered. Scientific alteration of the genetic code presents a different inquiry". Moreover, the Supreme Court expresses no opinion about such endeavours, limiting itself to hold that genes and the information they encode are not patent eligible "simply because they have been isolated from the surrounding genetic material". On the contrary, the so-called composite DNA, that is, exons-only strands of nucleotides created synthetically, can be patent eligible. The debate on barriers to sharing in the benefits of scientific development *and* its technological applications is far from being concluded.

When we say that this debate should involve all sectors of society, we acknowledge the very simple fact that these kinds of decisions build necessarily on a shared texture of ethical principles and values, distinguished from scientific and technical content as well as juridical aspects. Scientists and justices are not the only ones to which politicians (governments) should, and indeed do, refer. Thomas Pogge has shown how this awareness is deeply ingrained in the way we conceive and put into practice human rights. The effectiveness of a right to X relies not only on the legal system as a whole, but also on the socially widespread opinion that X is important. In this perspective "non-legal practices – such as a culture of solidarity among friends, relatives, neighbors and compatriots – may also play an important role" (Pogge 2008, p. 53). The standard of benefit sharing that we eventually come to consider as an obligation largely depends on this non-legal, social and cultural environment. Let's take the example of a country where the *legal* right to receive adequate health care, regardless of one's own economic condition, is not recognized. If a patient, who is suffering from a disease that could be easily treated, dies because he or she can simply not afford to pay for the treatment, the lack of a justiciable obligation could however be challenged by those people who continue to think that this is a blatant violation of a fundamental human right. The quest for reformation may trigger a powerful bottom-up dynamic that can be the premise also for a different legislation and a different institutional approach, strengthening, for example, the responsibility of the state to provide, in all cases, quality health care, at least for people who cannot afford it.

The same applies to the responsibility of shaping the rules of the intellectual property regime without making of them an insurmountable barrier preventing too many people from access to essential medicines and treatments. Pogge himself distinguishes a *push* and a *pull* approach. The former aims at selecting and financing potential innovators to start a specific research program under the condition that the potential benefits will be made available to competing pharmaceutical industries, rendering the innovation as widely accessible as possible, at the lowest market prices. The latter, on the contrary, turns to all potential innovators and is the solution proposed by Pogge: a pull plan does not invest in just some of them; leaves everyone free to develop the research they consider the most promising; and pledges to reward, through a special patent, the one which will produce a useful innovation, measured in terms of reduction of the global burden of disease. This is not the only idea on the table. Joseph Stiglitz, for instance, has also proposed an "innovation fund" to reward "those who make the really important discoveries", so that "drugs could be delivered (through generic producers) *at cost* to those suffering from disease" (Stiglitz 2006, p. 124). The crucial point that is worth underlining is that we really need a convergence of the efforts of many subjects, either in terms of a renewed and broader concept of solidarity, or, as a lesser alternative, a mutually satisfactory balance of interests. To use Pogge's words, his full-pull reform plan requires sound reasons to persuade peoples of wealthy countries and their representatives to support it and is pragmatically "based on the conviction that we will reach our common and imperative goal of universal access to essential medicines either in collaboration with the pharmaceutical industry or not at all" (Pogge 2008, p. 255).

Governments, citizens, economic and even for-profit actors are the most relevant players. The declinations of the concept of social responsibility may be therefore much different and complex. Therefore, a continuous activity of scrutiny is needed, as well as initiatives aiming at fostering exactly the idea of a shared commitment. The *Global Fund to fight AIDS, Tuberculosis and Malaria*, for example, was established as a private-public partnership exactly on this premise, bringing together "government bodies, international development partners (including United Nations agencies and donors), national civil society organizations (including local media, professional associations and faith-based institutions), the private sector, and communities living with or affected by the diseases". The idea of including "the voices of civil society as equal partners in all aspects", in particular, is made explicit as a key for the success of the initiative "at the global, regional and country level" (http://www.theglobalfund.org/en/about/partnership/).

6.4 Broadening the Idea of Benefit Sharing

If we look at two other well-known initiatives that have been undertaken in recent years, it is easy to find further evidence that the burden of inequality is two-tiered (or even three-tiered, considering also the regional level). *Grand Challenges in Global Health* focuses on 16 major challenges, "with the aim of engaging creative minds across scientific disciplines […] to work on solutions that could lead to breakthrough advances for those in the developing world" (http://www.grandchallenges.org/about/Pages/Overview.aspx). *The Reaching the Poor Program*, launched by the World Bank in cooperation with the Dutch and Swedish governments and, once again, with the Gates Foundation, points to the comparative disadvantage in the flow of benefits stemming from health, nutrition, and population programs, which can be traced back to just the condition of belonging to disadvantaged groups even within one and the same country, so that the poor among the poor come out to be double burdened. The figures related to the use of basic maternal and child health services, for instance, show that the coverage rates tend to be much higher among the best-off 20 % of developing countries' population than among the poorest 20 %, and even the distribution of benefits from government expenditure, in most cases undertaken precisely to help disadvantaged people, tends almost always to reach the poorest less frequently than it reaches the best-off (http://siteresources.worldbank.org/INTPAH/Resources/Reaching-the-Poor/summary.pdf).

The *Report* of the IBC on social responsibility and health has explicitly addressed these different faults of inequality, by assuming that nothing less than "the maximum of equality" should be considered as "the ultimate goal when everyone's right to life is at stake" and "the standard of equity and fairness must ensure in any case that the minimum to support human dignity be guaranteed to every individual". The normative content of this statement is in no way to be confused with a sort of double standard in terms of human rights. On the contrary, it assumes that the ultimate goal is one and the same and underlies therefore a thorough commitment to looking at viable strategies to approach that goal. At the domestic level, "the enjoyment of the highest attainable standard of health and access to quality health care without distinction – among others – of economic conditions, are obligatory goals for governments". In the transnational context, we have to come to terms with the evidence that "the States retain their freedom to choose what to do", so that non-legal practices, non-binding (at least not immediately) instruments, and the umbrella of other organizations become even more important "and a call for solidarity unavoidable" (UNESCO 2010, p. 23).

This is not only perfectly consistent with the mission of an organization which should strengthen the awareness that "since wars begin in the minds of men, it is in the minds of men that the defences of peace must be constructed". The call for solidarity is not a testament of wishful thinking. On the contrary, it is the premise of a very concrete and forward-looking education strategy, at the very crossroads of culture (cultures) and science.

It is undoubtedly necessary – to quote again the *Report* of the IBC – to work on a concept of solidarity much wider than the traditional group conception of it. Solidarity implies a "sense of belonging" which "can be a powerful motivating force" to turn "the passive acceptance of a common destiny to active work for common goals". Now, the problem is: how is it possible to harness solidarity "to goals such as the promotion of health and social development" and to "seeing health as a universal common good" (UNESCO 2010, p. 23)? The IBC is the most important global forum to address the new ethical issues raised by the development of biomedical sciences. This is why, recalling Durkheim's categories, it provides a privileged space to try and work out a concept of "organic solidarity" different from the one developed "within modern societies in which individuals are profoundly dependent on one another owing to the division of social labour" (Reichlin 2011, p. 369). We are probably not stretching the concept too much if we say that globalization processes have realized an organic yet non solidarist interconnection of all peoples of the world. What we are now called on to do is to mobilize cultural resources as well as establish education programmes aiming at fostering that "new form of organic solidarity that may arise only when human beings have acquired the consciousness of their ultimate universal connection, thanks to the scientific and technological developments that extend human agency in an unprecedented manner". This solidarity is "integral to the universalistic morality of justice" and based on the *reflection* "that human dignity is grounded on general features inherent in the human condition" (Reichlin 2011, p. 369). The general features that are at stake in the case of the human genome, but also, just to mention some other examples, in the various practices of exploitation and trafficking of human bodies and organs, as well as in the application of discoveries in fields such as bioengineering, neurosciences, nanotechnologies.

Networking is the key, but as long as it remains the networking of inequalities solidarity itself will remain and be perceived, at best, as a one-way flow and therefore a kind of beneficence rather than the evidence of the fundamental unity of humankind. The Bioethics Programme of UNESCO has anticipated this risk from the very beginning. Capacity building and education and awareness raising – together with providing an intellectual forum, standard-setting action and advisory role – are the main goals. Over these last decades, UNESCO offered a decisive contribution "to set up national ethics committees" and to capacity building at the national and regional level, "by facilitating the establishment of networks of institutions and specialists concerned with bioethics" and encouraging "the establishment or strengthening of regional bioethics information and documentation centres" (http://www.unesco.org/new/en/social-and-human-sciences/themes/bioethics/about-bioethics/). This work must go on and should perhaps be even more focused on that specific kind of benefit sharing which is mentioned in Article 15 of the *Declaration* of 2005, that is, capacity-building facilities for research purposes. It is worth reiterating that it is exactly in this perspective that solidarity is mentioned not only in Article 13, which is explicitly devoted to it, but also in Article 24, as a premise of the international cooperation which is required to promote the *Declaration* itself. States "should foster international dissemination of scientific information and encourage the free flow and sharing of scientific and technological knowledge". Within this framework, they should also promote cultural and scientific cooperation in order to enable developing countries "to build up their capacity to participate in generating and sharing scientific knowledge, the related know-how and the benefits thereof".

The *Recommendation on the Status of Scientific Researchers*, adopted by the General Conference of UNESCO in 1974, already urged Member States to "actively promote the interplay of ideas and information among scientific researchers throughout the world, which is vital to the healthy development of science and technology". As long as 'brain drain' remains the main outcome of young scientific researchers and physicians travelling abroad, the problem of the divide which keeps too many peoples in a condition of just *passive* sharing will not be overcome. Some countries, which were considered developing countries until a few years ago, have scaled the ranking of scientific production, in terms of number of patents and also in terms of capacity to attract high-skilled researchers. We need to improve this capacity everywhere, both because knowledge cannot remain a privilege for a few countries and because this capacity ensures that peoples establish their own agenda, without needing to import it from abroad. Of course, improving the production of something does not automatically entail its fair distribution, neither at the international nor at the domestic level. This is still a promising starting point. The most effective solidarity is always the one which sets the pillars of a two-way bridge and widens, step by step, the room for a cooperation entailing as little philanthropy as possible.

6.5 A Sustainable Holistic Approach

In 2002, the Johannesburg Summit, building on the vision which encompasses "social justice and the fight against poverty as key principles of development that is sustainable", that is, underlining that "solidarity, equity, partnership and cooperation were as crucial as scientific approaches to environmental protection" (UNESCO 2006, p. 9), proposed the Decade of Education for Sustainable Development. It was, so to say, a way to opt for a holistic approach with regard also to what was imposing itself as a breakthrough concept, including the objectives of the Millennium Development Goals and the Education for All Dakar Framework. In December of the same year, the United Nations General Assembly adopted resolution 57/254 to launch the Decade, which was planned from 2005 to 2014. UNESCO was entrusted with leading it and developing an Implementation Scheme.

It is no surprise to find out that health and quality of life were listed as bullet points, together with sustainable consumption, cultural diversity, water and energy, biosphere reserves and world heritage sites as places of learning, ESD in the knowledge society, citizen participation and good governance, poverty reduction, and intergenerational justice and ethics. The close entwining of health and environment, in particular, is clearly expressed: ill health "hampers economic and social development, triggering a vicious cycle that contributes to unsustainable resource use and environmental degradation. A healthy population and safe environments are important pre-conditions for sustainable development" (UNESCO 2006, pp. 41 and 19). I have already pointed out that the focus of the *Declaration* of 2005 on human rights does not exclude the full awareness of exactly this crucial interconnection between well-being of individuals and societies and a safe environment. Beyond Article 17, the underlining of the importance of biodiversity is already mentioned in Article 2 as one of the aims of the *Declaration*; the improvement of the environment is one of the goals of scientific progress that the article on social responsibility and health refers to; environmental conditions of special vulnerability are among the issues that, according to Article 24, are worth "special regard" in the context of international cooperation. Article 16 states the principle of protecting future generations and calls on to assess the impact of life sciences accordingly.

Nowadays, sustainable development is one of the overarching objectives of UNESCO and education for it has become a sort of umbrella-programme, further enriched through the reference to pivotal issues such as biodiversity, climate change, disaster risk reduction, gender equality, lifestyles, peace and human security, and urbanisation. It is not just the unprecedented extension of scientific and technological power in biosciences that is likely to cast a shadow of threat and uneasiness on our future. Ulrich Beck had already epitomized in the title of his successful book of 1986 (the *Risikogesellschaft*) an era that not only casts off traditional ways of life, but wrestles with the side effects of its own achievements. After two decades, he cannot dismiss the idea of a *world at risk*: "We are becoming members of a 'global community of threats'. The threats are no longer the internal affairs of particular countries and a country cannot deal with the threats alone. A new conflict dynamic of social inequalities is emerging" (Beck 2009, p. 8). The debate is also open on whether the necessity to grapple with this risky mixture of global threats and social inequalities should entail a deep reshuffling of the categories themselves on which the idea of human development has been long since predicated.

The *Declaration* of the 2008 Paris Conference on *Economic De-Growth For Ecological Sustainability And Social Equity*, for instance, renewed and updated the idea of "the limits of growth" worked out by the Club of Rome in its Report of 1972. Taking its cue from the seeming provocation of a post-development society (Latouche 1993), the *Declaration* calls for a paradigm shift "from the general and unlimited pursuit of economic growth to a concept of 'right-sizing' the global and national economies". Such a shift implies degrowth in wealthy parts of the world and increasing consumption, as quick as possible but *in a sustainable way*, by those in poverty in countries "where severe poverty remains". Aiming to meet "basic human needs and ensure a high quality of life, while reducing the ecological impact of the global economy to a sustainable level, equitably distributed between nations" (http://www.barcelona.degrowth.org/Paris-2008-Declaration.56.0.html), this proposal is also to be considered – and challenged – at the table where the ways to effectively implement a *global* social responsibility are discussed. The *Degrowth Declaration* of 2010 Barcelona Conference has reaffirmed the persuasion that "these proposals are not utopian" (http://www.barcelona.degrowth.org/Barcelona-2010-Declaration.119.0.html). In any case, this is the broad scope we have to address once we accept the idea that improving the standard of health that is really *attainable* for every human being is a core issue for bioethics, and the means to protect and improve health are a matter of society in addition to a matter of science.

6.6 Conclusion

More than 30 years ago, Halfdan Mahler, the Director General of the World Health Organization, wrote an article to explain the meaning of the target proposed for all Member States: health for all by the year 2000. An "imperative for change" was deemed the consequence of the observation of the few resources being invested in the health sector, of the unequal distribution of their benefits, of the continuous migration of physicians from poorer countries to richer ones (a persisting problem, which keeps hampering the efforts of many developing countries to set out an efficient health care system, especially Anglophone and Francophone countries), and the little control that ordinary people had over their own health care. It is easy to get the impression that everything we are talking about had already been said so many years ago. Health for all should be regarded as an objective of economic development and "not merely as one of the means of attaining it"; it demands, ultimately, "literacy for all"; it depends "on continued progress in medical care and public health", so that health services "be accessible to all"; it is "a holistic concept calling for efforts in agriculture, industry, education, housing, and communications" because "medical care alone cannot bring health to hungry people living in a hovel"; it implies a "new way of life" as well as the determination by governments "to promote the advancement of all citizens on a broad front" (Mahler 1981, pp. 6–7). The year 2000 has passed and the goal remains far from being accomplished. It is not a reason to give up. It is a reason to boost our understanding of global bioethics as *social* bioethics and act accordingly, both at the domestic and the international level. There is still much work to be done.

References

Beck, U. 2009. *World at Risk*. Cambridge/Malden: Polity Press. Translation from *Weltrisikogesellschaft*. Frankfurt a.M.: Suhrkamp (2007).
Dahl, R.A. 1971. *Polyarchy: Participation and opposition*. New Have: Yale University Press.
Freeman, R.A. 1994. The politics of stakeholder theory: Some future directions. *Business Ethics Quarterly* 4(1994): 393–408.
Latouche, S. 1993. *In the wake of the affluent society: An exploration of post-development*. London: Zed Books.
Mahler, H. 1981. The meaning of "health for all by the year 2000". *World Health Forum* 2(1): 5–22.
Martinez-Palomo, A. 2009. Article 14: Social responsibility and health. In *The UNESCO Universal Declaration on Bioethics and Human Rights*, ed. H. Ten Have and M. Jean, 219–230. Paris: UNESCO.
Pogge, T. 2008. *World poverty and human rights*. Cambridge: Polity Pres.
Reich, T.W. 1995. Introduction. In *Encyclopedia of bioethics*. New York: Macmillan.
Reichlin, M. 2011. The role of solidarity in social responsibility for health. *Medicine, Health Care and Philosophy* 14(4): 365–370.
Ricoeur, P. 2005. *The Course of Recognition*. Cambridge, MA: Harvard University Press. Translation from *Parcours de la reconnaissance*. Paris: Stock (2004).
Stiglitz, J. 2006. *Making globalization work. The next steps to global justice*. London: Allen Lane, Penguin Books.
UNESCO. 2006. *Framework for the UN DESD international implementation scheme*. Paris: Section for Education for Sustainable Development, Division for the Promotion of Quality Education.
UNESCO. 2010. *Report of the International Bioethics Committee of UNESCO (IBC) on social responsibility and health*. Paris: Division of Ethics of Science and Technology, Bioethics Section.

UNESCO. 2013. *The principle of respect for human vulnerability and personal integrity. Report of the International Bioethics Committee of UNESCO*. Paris: Division of Ethics of Science and Technology, Bioethics Programme.
United Nations Development Programme (UNDP). 1990. *Human development report 1990*. New York/Oxford: Oxford University Press.
WHO. 2008. *Closing the gap in a generation. Health equity through action on the social determinants of health*. Geneva.

Chapter 7
Ethics and Traditional Medicine

Emilio La Rosa Rodríguez

Abstract Traditional medicine is practiced in many countries and can contribute to improve human health provided that certain criteria are met such as, integration into the healthcare system, safety, efficacy, as well as quality. With regard to the ethical dimension of traditional medicine, the main difficulty lies in the diversity of ways in which it is practiced. The question currently being asked is whether traditional medicine is practicedin accordance with the bioethical principles of beneficence, autonomy and justice.

Beneficence is one of the pillars of medicine –providing beneficial treatment for the patient while avoiding or preventing harm– stems from a respect for life that must underpin any medical practice, including traditional medicine. Autonomy presupposes responsibility on the part of patients and their ability to make decisions for themselves. In the case of traditional practices, some therapies are based on magical or spiritual beliefs, however, free and informed consent must take account of the particular features of the proposed traditional treatment. The principle of justice is founded on the equal distribution of healthcare resources and expenditure, and the avoidance of discrimination. With regard to treatment, whether in modern or traditional medicine, the principle of justice requires that all patients with similar circumstances should have access to the same care, and that when resources are allocated to a group, the impact of this choice on others should be assessed. The same principle demands that all patients have access to effective and high-quality treatment, however, this cannot be achieved unless a concerted effort is made to assess the treatments available in modern and traditional medicine.

7.1 Introduction

Traditional medicine is a concept that goes beyond the scope of healthcare and touches social, religious, political and economic considerations. It is a set of systems for managing suffering that draw on theories relating to the body, health,

E. La Rosa Rodríguez, MD, PhD (✉)
Centre de Recherche et d'Etude Santé et Société, Paris 75017, France
e-mail: elarosa@free.fr

illness, suffering and healing that have their roots in the history of the cultures and religions that shape a country (Epelboin 2002). As defined by the World Health Organization (WHO), traditional medicine is "the sum total of the knowledge, skills, and practices based on the theories, beliefs, and experiences indigenous to different cultures, whether explicable or not, used in the maintenance of health as well as in the prevention, diagnosis, improvement or treatment of physical and mental illness" (WHO 2000). There are almost as many types of traditional medicine as there are cultures in the world. Traditional medical practices are the residue of generations and cultures that no longer exist, fragments of symbolic thought, rituals sacred and profane, which have been preserved, applied and passed on by "divine healers", also known as traditional practitioners.

Traditional medicine is not only practiced in developing countries, but also in industrialized ones, where it constitutes a rapidly expanding sector that draws on existing scientific and technological expertise. In such countries, traditional medicine is known as alternative, complementary or parallel medicine. Many methods are employed in traditional medicine, and these vary between individuals and societies. They range from the use of plants and animal organs to psychological and religious practices. Thanks to the combination of clinical and therapeutic knowledge, and clairvoyant techniques and abilities from invisible worlds, traditional practitioners give meaning to the ailments that affect the individual body and the social context of their patients by revealing the causality between events, whether biological or otherwise. People turn to traditional medicine chiefly because of its proximity and ease of access, as well as its availability and philosophical concordance with indigenous cultures. Traditional practices are expanding in many countries, and are thus taking on greater importance in healthcare as well as in economic terms. Yet one must not turn a blind eye to the inherent difficulties of traditional practices such as, lack of regulation, assessment, control and training, or to the dangers of offering healthcare that is most likely to be taken up by a specific sector of the population because it is accessible and cheap. In other words, conventional medicine might become the medicine of "the rich", and traditional medicine practice for the poor. Traditional medicine can make an effective contribution to improve human health provided that the countries in which it is practiced implement certain measures, such as those relating to the integration of traditional practices into the healthcare system and the efforts to address safety, efficacy, quality, accessibility and rational use. All these measures should be accompanied by regulatory and ethical standards. To this aim, in its work programme for 2010–2011, the UNESCO's International Bioethics Committee (IBC) included the subject of traditional medicine to elaborate the ethical implications of these widespread and highly varied practices. After two years work, as well as internal and external consultations with relevant stakeholders the IBC report on *Traditional medicine systems and their ethical implications* was developed in January 2013 (UNESCO-IBC Report 2013).

Finally, it should be emphasized that traditional knowledge of medicine deserves to be protected and recognized for its intrinsic value, particularly in social, spiritual, economic, intellectual, scientific, environmental, technological, educational and cultural terms. This collective heritage of indigenous peoples is a focal point for the

World Health Organization (WHO), which promotes the integration of traditional medicine into healthcare systems, and for the World Intellectual Property Organization (WIPO),[1] which endeavours primarily to "protect" traditional medical knowledge by means of intellectual property law.

7.2 Traditional Medicine Practices

With regard to the ethical dimension of traditional medicine, the main difficulty lies in the diversity of ways in which it is practiced.

Different types of traditional medicine have differing categorization of diseases, as well as differing diagnostic and therapeutic methods. Yet for the purpose of treatment they all make use of the following: medicinal plants, acupuncture and related techniques, chiropractic, manual therapies, "qi gong", "tai chi", yoga, naturopathy, other physical therapies and therapies based on magical and spiritual beliefs. These interventions have several features in common:

- therapeutic systems are adapted to the specific sociocultural and geographical environment and context that meet the healthcare needs of an ethnic group;
- they use local natural resources (plants, minerals, animals, water), in the prevention and treatment of illness, as well as elements closely associated with a culture and its belief system;
- in traditional medicine, health and illness are not binary or compartmentalized, but represent two ends of a spectrum (in keeping with the yin/yang dialectic) that is directly influenced by being in or out of balance with an environment that is understood as an expanded reality (physical environment, inhabited space, and symbolic space);
- they are inextricably linked to a culture and society. Just as health and illness are states that result from a balance or imbalance with the wider environment, so what constitutes "health" for one person in a specific situation could be deemed to constitute "illness" in other circumstances.
- therapeutic systems are natural and symbolic, and they draw on tradition for knowledge, organization, procedures and transmission. Nature and culture are a single entity and a dynamic reality in the majority of types of traditional medicine. Natural resources are necessary for life and are perceived as "partners" in the human experience of life.

---

[1] Negotiations are currently being held by the WIPO Intergovernmental Committee on Intellectual Property and Genetic Resources, Traditional Knowledge and Folklore (IGC) towards the development of an international legal instrument for the effective protection of traditional cultural expressions and expressions of folklore and traditional knowledge (including traditional medicine), and to address the intellectual property aspects of access to and benefit-sharing in genetic resources.

7.3 Ethical Considerations

Given the diversity of traditional medicine practices, it is important to promote an ethical approach that takes into account the universality of bioethical principles.

International organizations such as UNESCO and WHO, have tried to address ethical issues in traditional medicine. The UNESCO *Universal Declaration on Bioethics and Human Rights* (2005), has emphasized on the fair assessment of benefit and harm; respecting the principle of autonomy; informed consent; access to quality health care and benefit sharing, as the cornerstone of any practice that professes to be medicine (Articles 5, 6, 14 and 15).

The *Beijing Declaration*, which was the key outcome of WHO Congress on Traditional Medicine, for the first time offered international acknowledgement of the role of traditional medicine in health care. The Declaration clearly reaffirmed the need to regard these practices as something that "should be respected, preserved, promoted and communicated widely and appropriately based on the circumstances in each country" (Beijing Declaration 2008).

The World Health Organization (WHO 2002), considers that traditional medicine should be integrated into healthcare systems provided that it meets criteria for safety, efficacy, quality and rational use. Such integration requires a political framework that determines the place of traditional medicine in the healthcare system and establishes regulatory and legal mechanisms to govern training, research, best practice and equal access. However, in many countries there is a lack of standard for assessing traditional practice, therefore, the requirements of safety, efficacy and quality are important ethical concerns. Assessment of the products used by traditional medicine, for example medicinal plants, poses a number of ethical problems. Moreover, oversight and assessment of adverse events remain inadequate (WHO 2002). However, it should be noted that a number of long-standing traditional practices have demonstrated their therapeutic efficacy despite the absence of clinical trials and can be integrated into the therapeutic arsenal, as has occurred with a number of therapies in Western medicine (for example psychoanalysis and homeopathy).

The concept of rational use comprises several elements such as, qualification, authorization to practice, and correct use of a quality product. In order to address ethical concerns, it is therefore important to ensure that knowledge, qualifications and training are of good quality, and to create conditions that allow for the certification and authorization to exercise traditional practices. The question is, whether traditional medicine is subject to the principles of bioethics, namely; beneficence, autonomy and justice. Beneficence -providing beneficial treatment for the patient while avoiding or preventing harm (non-maleficence)- is one of the fundamental values in medicine. This principle is superior to that of "non-maleficence", as it implies active engagement in seeking a beneficial effect. Beneficence is a principle – or rather an imperative – that stems from a respect for life that must underpin any medical practice, including traditional medicine. To guarantee the principles of beneficence, the requirements of safety and effectiveness must be met. These requirements presuppose the establishment of adequate standards for training, certification, practice, assessment, control and research, as well as standards in the manufacture and sale of the products used in traditional medicine. Autonomy presupposes responsibility on the part of patients and their ability to make decisions for themselves. But in order to make decisions, patients must be informed, and this raises questions about the conditions in which informed consent is obtained. In the case of traditional practices, particularly therapies, which are based on magical or spiritual beliefs, free and informed consent should be taken into account seriously. In practice, it is very difficult to apply the principle of autonomy and informed consent in some type of traditional medicine such as spiritual therapies. The UNESCO-IBC report on traditional medicine states that "for spiritual therapies, the main problem is one of methodology, or epistemology, since it is very hard to take a scientific approach to a cultural phenomenon with psychosomatic effects". The report also emphasizes that, "these therapies reflect a given cultural reality and can be effective in that context. The problem is more the risk of misdiagnosis; if and when the intention is to cure an organic pathology with unsuitable treatment, the consequences can be serious for the patient" (UNESCO-IBC Report 2013). However, this should not result in discarding these principles in this or other types of traditional medicine practice. Rather, it is important to overcome these difficulties by attempting to determine and implement ethically sound procedures that are sensitive to cultural considerations. One possible solution is to ensure the close and sustainable involvement of the community in finding adequate solutions, and the participation of the community in research into traditional medicine. As stated in the UNESCO Report on traditional medicine, "…traditional medicine comes in all shapes and sizes. Nor is it limited to diagnosis and treatment. It implies a specific approach to life, death, health and illness. It entails a different view of the patient, practitioner, patient-practitioner relationship, health services, risk factors, etc. On the other hand, where medical practice is an integral part of a group's culture, where the patient is a "social fact" whose treatment involves the group, and where the traditional practitioner plays a key role in the community, application of the principles of autonomy, individual responsibility and consent presents a challenge. This can be met only by critical recognition of beliefs and traditions" (UNESCO-IBC Report 2013).

It should be noted that research in the field of traditional medicine must be as rigorous as that of modern medicine and must adhere to similar standards, especially given the substantial risk that individuals could be recruited for studies by exploiting their beliefs. National regulations regarding research into traditional medicine must make provision for the protection of such individuals as well as the protection of the inherent characteristics of traditional therapies. However, some researchers take the view that traditional medicines tested on thousands of people over decades or centuries should not be subject to the same assessment procedures as new drugs (Tilburt and Kaptchuk 2008). To act in a free and informed manner means acting intentionally with an understanding of the risks involved, without being controlled by any influences that might determine an individual's choices.

The principle of justice is founded on the equal distribution of healthcare resources and expenditures, as well as the avoidance of discrimination. In terms of care, whether in modern or traditional medicine, the principle of justice requires that all patients with similar circumstances should have access to the same care, and that when resources are allocated to a group, the impact of this choice on others should be assessed. The same principle demands that all patients have access to effective and high-quality treatment, and this cannot be achieved unless a concerted effort is made to assess the treatments available in modern and traditional medicine.

Traditional medicine must not be an alternative for the poor, nor should it be a pretext for failing to improve access to the best diagnostic techniques and treatment. Moreover, respect for tradition and cultural identity should not be used as a reason for reducing access to effective care or withholding proper information about health conditions. Everyone should have access to high-quality care, whichever type of medicine they use. It would be unacceptable to allow a two-tier healthcare system to develop, one that is easy to access and inexpensive for those on modest incomes and another for the rich; this would promote discrimination. Traditional and modern medicine can coexist provided that bridges are built between them. When a traditional treatment proves to be effective, it should be made available to all; this would require modern and traditional medicine to be complementary to each other. However, respect for cultural diversity and the fact that poorer people will not seek out something of which they are unaware, can only lead to disengagement and a slackening of the principles of justice and solidarity in a world in which life expectancy is over 80 years for some people and less than 50 years for others. In this respect, public health systems are charged with the responsibility for ensuring that the right to high-quality care is respected (UNESCO-IBC Report 2013).

7.4 The UNESCO IBC Report on Traditional Medicine

Traditional medicine rests on analogical reasoning and a holistic approach to health and disease, while modern medicine is evidence-based and built on scientific knowledge. In its report on *Traditional Medicine Systems and Their Ethical Implications* (2013), the UNESCO-IBC examined a wide range of traditional medicine around the world and elaborated several benefits and advantages of traditional medicine such as: availability and affordability; cultural acceptability:, holistic and person-centered approach; protection of biodiversity.

By raising several ethical concerns in the application of traditional medicine, such as, autonomy, safety, efficacy, quality of products, non-discrimination and biopiracy, the report suggests a set of guidelines for actions. First, the report suggests that traditional medicine must be taken seriously as a branch of medicine and therefore like modern medicine, to ensure the safety, efficacy and quality of traditional medicine it requires assessment and monitoring. The integration of traditional medicine into healthcare system and regulating the practice of traditional medicine have also emphasized in this guidelines.

The guidelines has emphasized on education and training for both physicians and traditional medicine practitioner. It suggests that physicians should learn about the cultures of indigenous peoples and respect their beliefs and customs and also traditional practitioners should undergo appropriate training to be able to work with and effectively complement the national health system. The guidelines call for freedom of choice in using traditional therapies. It suggests that by improving basic services and facilities, education and the pursuit of excellence in local scientific communities, traditional practitioners, governments and international institutions can help to ensure that traditional practice is a genuine choice across the globe. By acknowledging the importance of protecting developing countries against the risk of exploitation of their traditional knowledge and natural resources, the guidelines urges researchers and multinational companies wanting access to these resources to obtain prior informed consent and to share the results and benefits (UNESCO-IBC Report 2013).

7.5 Conclusion

The criticisms most often made of traditional medicine emphasize its current inability to demonstrate therapeutic efficacy in scientific terms. However, over the centuries a number of traditional therapies have demonstrated their efficacy in practice, which is why they have survived. This does not exonerate traditional medicine from the need to improve supervision, training, certification, assessment, controls and research in order to make it more effective and in order to ensure that it can then be integrated into healthcare systems. Unfortunately, such efforts to improve traditional medicine have only been formalized in a very small minority of countries. Pending further progress in these areas it is necessary to protect traditional medicine beyond intellectual property aspects, and our focus should be on respecting the vast quantities of knowledge and thought that have been accumulated by indigenous populations over millennia. Equally important is a respect for the vibrant and authentic heritage of ancient medicines, which have plenty to teach us. There must also be respect for medicinal plants and the communities which cultivate them The development of new medicines that derive their active ingredients from plants will only be possible if communities are reassured that they will benefit from the dividends of the refinement process. Indigenous rights must be protected and measures put in place to guard against the commercial misuse of plants by establishing legal mechanisms governing their distribution.

In this regard, Article 31 of the *United Nations Declaration on the Rights of Indigenous Peoples* (United Nations 2007)[2] should guide efforts at the national and

---

[2] "Indigenous peoples have the right to maintain, control, protect and develop their cultural heritage, traditional knowledge and traditional cultural expressions, as well as the manifestations of their sciences, technologies and cultures, including human and genetic resources, seeds, medicines, knowledge of the properties of fauna and flora, oral traditions, literatures, designs, sports and traditional games and visual and performing arts. They also have the right to maintain, control, protect and develop their intellectual property over such cultural heritage, traditional knowledge, and traditional cultural expressions. In conjunction with indigenous peoples, States shall take effective measures to recognize and protect the exercise of these rights".

international levels to protect traditional medical knowledge, and should form the basis of a legally binding international instrument.

References

Beijing Declaration. 2008. WHO congress on traditional medicine. Available at: http://www.who.int/medicines/areas/traditional/TRM_BeijingDeclarationEN.pdf

Epelboin, A. 2002. Médecine traditionnelle et coopération internationale. *Bulletin AMADES* 50. Available at: http://amades.revues.org/900

UNESCO Report. 2013. *International Bioethics Committee report on traditional medicine systems and their ethical implications*. Paris: UNESCO.

United Nations. 2007. United Nations Declaration on the Rights of Indigenous Peoples. Resolution adopted by the General Assembly, 13 Sept 2007. Available at: http://www.un.org/esa/socdev/unpfii/documents/DRIPS_en.pdf

Tilburt, J.C., and T.J. Kaptchuk. 2008. Herbal medicine research and global health: An ethical analysis. *Bulletin of the World Health Organization* 86: 577–656.

World Health Organization. 2000. *General guidelines for methodologies on research and evaluation of traditional medicine*. WHO/EDM/TRM/2000.1, Geneva.

World Health Organization. 2002. *WHO traditional medicine strategy 2002–2005*. Geneva: World Health Organization.

Chapter 8
Biobanks: Balancing Benefits and Risks

Ewa Bartnik and Eero Vuorio

Abstract Since its creation in 1993, the International Bioethics Committee (IBC) of UNESCO has been actively involved both in analysing bioethical problems and in proposing related guidelines. Currently IBC is focusing on a number of bioethical problems in the rapidly changing world of biomedical research. The purpose of this chapter is to review the progress in biobanking of human specimens and their high-throughput analysis into data. Biobanks are becoming repositories of human genetic material and data and thereby play an important role in the advancement of human health and in research and development in life sciences and biomedicine. More importantly, systematic collection of human samples and data provides the basis for better stratification of diseases, for development of personalized medicine and for development of health policies throughout the world. However, availability of bio-banked samples and derived genetic data may also create problems concerning informed consent, incidental (unsolicited) findings and privacy, which have also been discussed by IBC.

8.1 Introduction

The International Bioethics Committee (IBC) has been active in evaluating bioethical problems which have arisen for the past 20 years and in proposing bioethical guidelines. This period has seen unprecedented progress in the field of biomedical sciences, associated with both hopes for the cure of many diseases and with certain fears concerning the pervasiveness and the ease of accessing very personal

---

E. Bartnik (✉)
Institute of Genetics and Biotechnology, Faculty of Biology,
University of Warsaw, Warsaw, Poland

Institute of Biochemistry and Biophysics, Polish Academy of Sciences,
Warsaw, Poland
e-mail: ebartnik@igib.uw.edu.pl

E. Vuorio
Molecular Biology, University of Turku, Turku, Finland

information about each individual. Progress in high-throughput techniques has made it possible to obtain huge amounts of medically relevant data from patients and healthy subjects, and together with equally fast development in medical informatics is a vivid example of the speed at which the IBC needs to follow the global bioethical environment. This also means that documents produced ten years ago, such as the *International Declaration on Human Genetic Data*, need to be revisited for their content as they do not properly address biobanking.

The working program of the IBC for 2012–2013 has focused on various areas of biology and medicine where discrimination or stigmatization may become a problem, due to both existing and emergent technologies, one of the latter being biobanks. The aim of this chapter is to discuss some ethical concerns and the role that biobanks play in the progress of contemporary medicine and genetics as examples of the types of problems arising in society with the rapid development of new technologies associated with analysis of human health and their consequences.

8.2 Paradigm Shift in Medicine

The paradigm of modern medicine is shifting rapidly towards improved diagnostics and proper classification (stratification) of diseases accompanied by insightful analysis of DNA of individual patients. The aim is personalized medical treatment, perfectly suited to the disease and the patient. This encompasses molecular diagnosis of the patient's disease, but also ascertaining of individual differences affecting such parameters as drug metabolism, disease course and response to treatments. Also environmental and occupational exposure, nutrition, and life style can be taken into account. In short, this type of approach is expected to expedite diagnosis of diseases, help doctors select the best treatment and avoid using drugs which are known not to be effective, or may even be harmful, to the specific variant of the disease the patient has. In fact, prevention of harmful side effects is an important driver of personalized medicine. Another driver of the process is the pharmaceutical industry, which has realized that even the best drugs are only effective when given to the right patients for whom they have been designed. In the long run, the increased efficiency of personalized treatments is expected to result in considerable savings to the health care system and major benefits for the patients. While there is no empirical study to clearly demonstrate the former expectation, there is plenty of evidence for major benefits of personalized treatments for the patients as demonstrated by greatly improved survival rates of patients with several different types of cancers.

Some positive effects of personalized medicine can already be observed for many diseases, including for instance breast cancer, where earlier detection and better testing and more individualized treatment have led to improved survival of patients. Screening of new-borns for a number of diseases whose early detection leads to either prevention of the disease or improvement in the quality of life is another example where the benefits are obvious.

Although the effects of personalized genome analysis are apparent in the field of diagnosis and treatment of an increasing number of diseases, the enthusiasm for determining what is hidden in our genome is not without problems and may still outweigh the real benefits. This is definitely so in the case of most "common" diseases which are a complex outcome of genetic predispositions and environmental effects. In most cases intensive research has identified many genes each of which is responsible for only a very small part of the variability of the analysed disease or trait, and altogether they do not even begin to account for it. Nevertheless, as sequencing is becoming cheaper people are beginning to see genome analysis as a sort of investment in their own health. Currently whole genome analysis is rarely available in the clinical setting, though it is increasingly being used, either in the form of whole genome sequencing or as exome sequencing -analysis of only less than 2 % of human DNA which codes for proteins- for diagnosis of diseases which are believed to have a genetic cause which is not found by standard tests (Lupsky et al. 2010, 2013). Genome sequencing may also be increasingly used in the future in following the effects of cancer therapy or determining possible drugs which can be used for a given patient and stage of the disease (Vogelstein et al. 2013).

8.3 A Long History of Disease-Oriented Biobanks

Research on the diseases caused by multiple genetic and environmental factors has led to the increasing need for very large groups of patients (and controls) whose genomes are scanned for mutations or polymorphisms which contribute to the analysed diseases or traits. This has led to the increased popularity of large collections of samples and related research data, known as biobanks. As defined in the *Report of the IBC on the Principle of Non-discrimination and Non-stigmatization*, "the term biobank refers to collections of biological material (blood, tissue, DNA etc.), whether collected as part of routine health care or as research-oriented cohort studies". Although the word biobank is a relatively new invention, systematic collection of patient-derived samples, i.e. disease-oriented biobanks, is an old institution. Health care legislation of many countries considers routine collection and storage of human samples, originally obtained for diagnostic purposes, as an obligatory part of the quality control system of health care. However, when combined with different health-related registries, such biobanks can reveal important information on individual patients, and provide much-needed information on efficacy of different treatment modalities. Long-term storage of samples is important as this makes it possible to follow the progression, remission or relapse of a patient's disease, its response to treatment and allows reclassification of the disease when molecular diagnostic tools are developed (UNESCO Report 2014). While storage of biological samples and related health information has rarely been considered a serious ethical problem when part of routine health care, use of biobanked samples for research purposes that could not be foreseen at the time of sample collection can raise ethical concerns. These often stem from determination of an individual DNA sequence from a small tissue sample, drop of blood or even from small numbers of cells each individual leaves just like fingerprints. However, individual DNA can even be identified in complex mixtures (Homer et al. 2008). Since the determination of the first complete human DNA sequences at the turn of the millennium, the cost of sequencing has dropped dramatically (by several orders of magnitude) and the speed has increased equally. Such technological development could not have been predicted 10–15 years ago. The ethical challenges that this new information brings about are the predictive nature of DNA sequence as discussed in this chapter, as well as the fact that DNA sequence is probably the most unique identifier of any human being. This is the basis for a special attitude towards the DNA sequence, even among those who do not worry about using biometric passports at border controls and giving their fingerprints for immigration officials.

8.4 Population Biobanks: The Power of Many

Although collections of different types of materials from various organisms have been with us for a long time, two developments have opened up new possibilities, for both vastly improved understanding of human diseases and for improper use of knowledge relating to individuals. The first is the progress in sequencing technology, discussed above. The second is the creation of numerous, often very large population-based biobanks as long-term repositories of human materials within projects aimed at resolving the causes of complex diseases or at finding biomarkers which could allow their early detection. Such biobanks were first started in Northern European countries in the 1960s, where public support for this type of research and public trust to the researchers studying biobanked samples has been the highest. Currently, large well-known biobanks exist in the United Kingdom, Iceland, Estonia, the Faroe Islands and the Nordic Countries, but also elsewhere. In Europe, a Pan-European Biobanking and BioMolecular Resources Research Infrastructure (BBMRI) has been established to network European biobanks containing samples of tens of millions of individuals (Pan-European Biobanking 2014).

Population biobanks face technical challenges of ensuring proper coding, storage and access for sometimes many years, and the bioethical problem of the extremely complicated informed consent in these cases. IBC has prepared a special report on informed consent and it is also addressed in the UNESCO Declarations on Human Genetic Data[1] and on Bioethics and Human Rights,[2] (UNESCO Declaration 2003 and 2005) but when these documents were prepared the problem of informed consent either given in a very broad form (as it is impossible to predict what dis-

---

[1] International Declaration on Human Genetic Data (2003): A number of provisions deal with the issue of consent related to the specific subject of human genetic data and further develop the provisions of the Universal Declaration on the Human Genome and Human Rights (1997) on this issue. Article 8 deals with consent to the collection of biological samples and human genetic data, Article 9 is devoted to the withdrawal of consent and Article 10 addresses the issue of the right to decide whether or not to be informed about research results.

[2] Universal Declaration on Bioethics and Human Rights (2005): Devoted two articles to the issue of consent – Article 6 addresses the principle of consent and Article 7 covers the case of persons without the capacity to consent.

eases will be analysed in 20 or so years) or the need to re-contact the biobank donors each time some new testing procedure is to be implemented was not addressed. These issues have been extensively discussed in many documents and publications, and will not be analysed further here (Hansson 2009; Hoeyer 2008).

In some Northern European countries, new biobank legislation considers broad informed consent, including reuse of old sample collections, perfectly fine when combined with a possibility to opt out. The trend is clearly towards individuals using their right to participate in medical research and to benefit from the results of such research, as stated in the UN Declaration of Human Rights. This indicates that some countries have found an acceptable way to deviate from the core commitments of informed consent by other methods of informing the citizens, performing a rigorous scientific and ethical review and assuring the highest possible level of data protection.

8.5 The Problem of Incidental Findings

A problem which has arisen in relation to biobanks but also with non-biobank related genetic testing concerns what should be done with results which may affect the health of the analysed person, but which are incidental, that is they were not the object of the original research plan (Wolf et al. 2012). The UNESCO declarations, for instance the *International Declaration on Human Genetic Data*, Article 10, have stressed the right of the patient not to know the results of his or her genetic test. However, it has been commonly accepted that tests should not be performed on minors if the disease that is tested for cannot be prevented or does not appear until a later age; but should be deferred until the child has reached an appropriate age to decide if he or she wishes to be tested. These principles have recently been ignored by the American College of Medical Genetics and Genomics as they have stated that patients undergoing genomic sequencing should be informed about the results concerning 57 genes which could lead to a disease or serious problems in the future, even if they do not wish to be informed and even if this information concerns minors (ACMG Recommendations 2013). This problem of incidental – or unsolicited– findings has also been extensively analysed by the European Society of Human Genetics (van El et al. 2013) and an interesting proposal of a qualified disclosure policy where a "menu of options" is offered to the participants has been made (Bredenord et al. 2011).

There is probably no way to stop the expansion of genetic services being offered, whether through participation in biobank projects, commercial companies or any other methods, and everyone is entitled to the access to the best possible diagnosis and treatment. The problem is now to a large extent, understanding of the exact nature of the research proposal and the possible outcomes, and this has become very complicated. The difference between "certainty" and "risk determination" in biological diagnosis is complex even for simple monogenic diseases. For instance a healthy person carrying a mutated Huntington gene can be informed that he will certainly succumb to Huntington's disease at the average age of 40, but he can become ill earlier or later. A person carrying polymorphisms that increase the risk of diabetes can only be informed that his or her risk is increased, and even that may not be accurate until all genes predisposing to diabetes are known and the environmental influence is properly taken into consideration.

The concept of genetic risk in modern medicine has been subject to a number of publications (Rørtveit and Strand 2001; Gottweis 2005) dealing philosophically with risk and different modalities of uncertainty as well as ignorance. One of the aims of biobanking and widespread genetic analyses is to strengthen the evidence base for genetic risk in medical practice.

In spite of hearsay stories that some people would love to do prenatal sequencing of their child's DNA, we are not at that stage yet. However, a lot of people prematurely believe that genome analysis can be directly applied to their benefit, whereas currently this is only true for diseases caused by mutations in single genes. The other side of the coin is fear of finding out something that may require expensive treatment, affect health insurance and possibly lead to discrimination in employment and to stigmatization. This may lead to avoidance of genetic screening programs, refusal to donate samples for biobanks, and thus not participating in the tremendous possibilities of modern medicine. Depending on the cultural and societal environment, both participation and non-participation may lead to discrimination and/or stigmatization.

8.6 International Declaration on Human Genetic Data

The *International Declaration on Human Genetic Data* was adopted by the UNESCO General Assembly in 2003 and is pertinent to a lot of the current problems, though it has always remained in the shadow of the two Universal UNESCO bioethical declarations on the Human Genome and on Bioethics and Human Rights. This Declaration has pointed out the sensitivity of genomic and proteomic data, indicated guidelines for the options of persons whose genetic material was being tested and gave firm guidelines for the conditions for which minors and in general persons incapable of expressing their consent should be tested. It should be noted that the above mentioned proposal by the American College of Medical Genetics and Genomics runs counter to several points of the *International Declaration on Human Genetic data*, namely to Art. 10 concerning the right to decide whether or not to be informed about research results. However, to some extent it is in conflict with Art. 8(d) which states that, "screening and testing of minors and adults not able to consent will normally be only ethically acceptable when it has important implications for the health of the person and has regard to his or her best interest". This is addressed in more detail by the *Universal Declaration on Bioethics and Human Rights* which states that (Article 7 (b), "Research which does not have potential direct health benefit should only be undertaken by way of exception". This is in general taken to mean that tests for diseases in persons unable to consent should be only performed if the results benefit them directly and is in general interpreted as not analysing genes for diseases such as hereditary breast cancer or Huntington's disease which will only manifest in adult age. The idea that people should be contacted and informed about incidental findings will not go away, there are some excellent papers describing when and how such results should be communicated, but perhaps the crucial description is that they should be "clinically actionable" (Wolf et al. 2012). This means that something can be done to prevent or alleviate the course of the disease. Of course a possible solution is the one mentioned above namely the "menu of options" pertaining to disclosure of the incidental findings to the tested individual. On the other hand there is such a thing as the "right not to know", and this should not be infringed on the basis of paternalistic assumptions that such information will benefit the patient, whether he wants it or not.

8.7 The Way Forward

As mentioned in the beginning, we believe that rapid technological development together with obvious benefits to individuals has brought biobanks to modern biomedical research and health care. The IBC and other international, regional and national ethical bodies will obviously have to follow and monitor the development of the field constantly. However, the proper education of individuals and societies is important to minimize the risks of genetic diagnostics and biobanking. Studies conducted in Europe demonstrate (Eurobarometer; Gottweis et al. 2011), that the public's understanding of biobanks and genetics in general is often limited. The studies also show that increased understanding of biobanks increases the acceptability of biobank-based research by individuals. In this regard, the understanding of genetic risk (of developing a disease) is even more challenging.

Low level of genetic literacy has been recognized as a problem by professional organizations, governments, health care providers and patient organizations. A quick search of the World Wide Web shows a range of information aiming to clarify what genetic risk, predisposition etc. mean and also address the ELSI (ethical, legal and societal) issues of genetic research. These include e.g. the Human Genome Project www-site of the National Institutes for Health in the US (http://www.nhgri.nih.gov) and web-sites of prestigious journals like Nature (http://www.nature.com/genomics). All language groups we have checked seem to have similar kind of information available in the national language(s) of the population. A point has been made that genetics education should begin already at school (Kung and Gelbart 2012). Unfortunately, the web also contains a large number of advertisements for commercial DNA-based nutrition, fitness and training programs without scientific support which does not make the ordinary citizen's life any easier and re-emphasizes the importance of genetic literacy.

Another success factor in ongoing biobanking projects (e.g. in the Faroe Islands), is public trust in health authorities, a social health care system and excellent electronic health records which help the people get maximal benefit from research data.

There is no doubt that like other medical information, results from genetic testing must be considered confidential, in which under normal practice, the doctor-patient relationship protects against disclosure of genetic information.

Biobanking and genetic research on indigenous populations is an especially sensitive issue (Taniguchi et al. 2012). One of the concerns is that the benefits from such research may not reach the community which is being studied. In some cases, e.g. in Australia, this has created a barrier preventing genetic research, which can also be viewed as a discrimination. In North America some web pages run by indigenous populations encourage them to stay away from modern genetic testing and stick to their traditional ways of life. The IBC has a lot of work ahead of it to accommodate all these ethical concerns in current and future guidelines.

8.8 Future Challenges

The extremely rapid decrease in the cost of DNA sequencing is making the idea of universal genome (or perhaps only exome) sequencing available for an increasing number of individuals. However, this will not resolve the current problems of incidental findings. We predict that patient- and biobank-derived genetic information will gradually become part of routine health records in countries with developed public health care systems. Genetic data will then be treated as part of other confidential health information. Such development will also put all incidental findings into a clinical perspective where the doctor in charge will convey this information to the patient as appropriate. It may be worthwhile to end the chapter with a quotation from the latest IBC document, *The Report of The IBC on the Principle of Non-discrimination and Non-stigmatization*, which points out the need for education at many levels to help avoid ethical problems related with biobanking: "As molecular analysis of biobanked human samples is becoming increasingly important for modern medical practice and research and genetic testing has become accessible to lay people, there is a burning need to educate individuals and populations how to interpret the results of genetic analyses, including incidental findings. We recommend that governments and medical professionals initiate educational programs to inform their citizens of the biobanking, new molecular classification (stratification) of diseases and subsequent development of targeted therapies, i.e. personalized health care. Genetic counselling must be made available to complement educational programs in order to prevent stigmatization of individuals carrying gene alleles causing or predisposing to disease. We also recommend legislative measures to prevent discrimination based on genetic testing when seeking employment or health/life insurance. With respect to this goal, the role of the media in disseminating knowledge and fostering the awareness of the new challenges to address is also a fundamental one".

References

American College of Medical Genetics and Genomics. 2013. ACMG recommendations for reporting of incidental findings in clinical exome and genome sequencing. Available at: https://www.acmg.net/docs/ACMG_Releases_HighlyAnticipated_Recommendations_on_Incidental_Findings_in_Clinical_Exome_and_Genome_Sequencing.pdf. Last visited 8 Mar 2014.

Bredenord, A.L., N.C. Onland-Moret, and J.M. Van Delden. 2011. Feedback of individual research results to research participants: In favour of a qualified disclosure policy. *Human Mutation* 32: 861–867.

Gottweis, H. 2005. Governing genomics in the 21st centrury: *New Genetics and Society* 24: 175–193.

Gottweis, H., et al. 2011. Connecting the public with biobank research: Reciprocity matters. *Nature Reviews Genetics* 12: 738–739.

Hansson, M.G. 2009. Ethics and biobanks. *British Journal of Cancer* 100: 8–12.

Hoeyer, K. 2008. The ethics of research biobanking: A critical review of the literature. *Biotechnology and Genetic Engineering Reviews* 25: 429–452.

Homer, N., et al. 2008. Resolving individuals contributing trace amounts of DNA to highly complex mixtures using high density SNP genotyping microarrays. *PLoS Genetics* 4, e1000167.

Kung, J.T., and M.E. Gelbart. 2012. Getting a head start: The importance of personal genetics education in high schools. *Yale Journal of Biology and Medicine* 85: 87–92.

Lupsky, J., et al. 2010. Whole-genome sequencing in a patient with Charcot-Marie-Tooth neuropathy. *New England Journal of Medicine* 3629: 1181–1191.

Lupsky, J., et al. 2013. Exome sequencing resolves apparent incidental findings and reveals further complexity of SH3TC2 alleles causing Charcot-Marie-Tooth neuropathy. *Genome Medicine* 5: 57.

OECD Guidelines on Human Biobanks and Genetic Research Databases. 2009. Available at: http://www.oecd.org/science/biotech/44054609.pdf. Last visited 8 Mar 2014.

Pan-European Biobanking and BioMolecular Resources Research Infrastructure. 2014, Available at: www.bbmri-eric.eu. Last visited 8 Mar 2014.

Rørtveit, G., and R. Strand. 2001. Risk, uncertainty and ignorance in medicine. *Tidsskrift for den Norske Lægeforening* 121: 1382–1386.

Taniguchi, N.K., et al. 2012. A comparative analysis of indigenous research guidelines to inform genetic research in indigenous communities. *The International Indigenous Policy Journal* 3(1). Available at: http://ir.lib.uwo.ca/iipj/vol3/iss1/6.

TNS Opinion and Social, "Biotechnology", Special Eurobarometer 341/Wave 73.1, Report for European Commission (October 2010).

UNESCO. 1997. Universal Declaration on the Human Genome and Human Rights. http://www.unesco.org/new/en/social-and-human-sciences/themes/bioethics/human-genome-and-human-rights/. Last visited 8 Mar 2014.

UNESCO. 2003. International Declaration on Human Genetic Data. Available at: http://www.unesco.org/new/en/social-and-human-sciences/themes/bioethics/human-genetic-data/. Last visited 8 Mar 2014.

UNESCO. 2005. Universal Declaration on Bioethics and Human Rights. Available at: http://www.unesco.org/new/en/social-and-human-sciences/themes/bioethics/bioethics-and-human-rights/. Last visited 8 Mar 2014.

UNESCO. 2008. Report of the International Bioethics Committee of UNESCO on Consent. Available at: http://unesdoc.unesco.org/images/0017/001781/178124e.pdf. Last visited 8 Mar 2014.

van El, C.G., et al. 2013. Whole genome sequencing in health care. Recommendations of the European Society of Human Genetics. *European Journal of Human Genetics* 21: S1–S5.

Vogelstein, B., et al. 2013. Cancer genome landscapes. *Science* 339(6127): 1546–1558.

Wolf, S.M., et al. 2012. Managing incidental findings and research results in genomic research involving biobanks and archived data sets. *Genetics in Medicine* 14: 361–384.

Chapter 9
The Risk of Discrimination and Stigmatization in Organ Transplantation and Trafficking

Alireza Bagheri

Abstract The global shortage of organs for transplantation has led to unethical practices in organ transplantation, such as organ commercialism and trafficking. Concerns have been raised about unjust and discriminatory allocation of the available organs in organ transplant programs as well as exploitation and stigmatization of individuals who provide their organs through organ trafficking and tourism. There have been global efforts to describe unethical practices in organ transplantation and in tackling organ commercialism and trafficking, international documents have justified their arguments mostly based on the exploitation inherent in organ sales and trafficking. Missing in the discussion of organ transplantation and trafficking are the perspectives of vulnerable patients as organ recipients and poor people as organ providers, and the discrimination and stigmatization they experience.

This chapter elaborates the risk of discrimination and stigmatization in organ transplantation and trafficking, and reviews current global efforts against unethical practice in organ transplantation, including the recent UNESCO report on non-discrimination and non-stigmatization. It calls all stakeholders to ensure that in the process of organ transplantation, organ donors and recipients are not subject to discrimination and stigmatization.

9.1 Introduction

Life-saving transplant technology has grown dramatically; however, there is a shortage of human organs and this is not limited to solid organs; there is an increasing demand for cell and tissue transplantation as well. In the current situation, the demand for transplants has grown far faster than the supply of available organs. In order to overcome the shortage of organs, several proposals have been introduced to expand the donor pool (Bagheri 2007). However, the controversial practice of organ markets and commercialism to address the local need for organ transplantation, and

A. Bagheri (✉)
Tehran University of Medical Sciences, Tehran, Iran
e-mail: bagheria@tums.ac.ir

the ethically condemnable practice of organ trafficking to provide organs for international patients have been subject to critical examinations. In fact both practices cause exploitation, discrimination and stigmatization of organ providers and/or organ recipients (UNESCO-IBC Report 2014). Although, discrimination and stigmatization may occur in all areas of healthcare and medical research, the focus of this chapter is to explore how individuals may become subject to discrimination and/or stigmatization in the application of transplant technology. Over the last two decades, international organizations such as the World Health Organization (WHO 2010) and professional associations have developed guidelines and recommendations to tackle organ commercialism and trafficking, however, none of them have examined the issue in the context of discrimination and stigmatization in adequate detail. It is important to thoroughly explore how the application of transplant technology may cause stigmatization and/or discrimination in a legitimate procurement program as well as through organ trafficking and tourism. The recent *UNESCO-IBC Report on The Principle of Non-stigmatization and Non-discrimination* has focused on this issue and if its recommendations are implemented, it will be very instrumental in preventing stigmatization and discrimination in organ transplantation.

9.2 Organ Transplantation and Trafficking: Current Global Initiatives

For more than two decades, international organizations and governments around the world have recognized the global problem of organ trafficking and the exploitation of poor people as organ providers around the world. There are several major international initiatives to regulate organ transplantation and to tackle organ trafficking (Bagheri and Delmonico 2013).

Since 1987, the World Health Organization (WHO) became involved in regulating organ transplantation when the World Health Assembly expressed concern about the commercial trade in human organs. In 1991, the document, *WHO Guiding Principles on Human Organ Transplantation* was endorsed by the member states. Later, after a long process of extensive consultations at national, regional and sub-regional levels with all stakeholders, this document was revised and renamed the *WHO Guiding Principles on Human Cells, Tissues and Organ Transplantation*, and was endorsed in May 2010. The *Guiding Principles* were intended to provide an orderly, ethical and acceptable framework for the acquisition and transplantation of human cells, tissues and organs for therapeutic purposes. The document introduced 11 guiding principles which emphasize the need for: organ donation without any monetary payment or other reward of monetary value; protection of personal anonymity, and privacy of donors and recipients; donor's informed and voluntary consent for live donation; and allocation of organs, cells and tissues based on clinical criteria and ethical norms, not financial or other considerations (WHO 2010).

In another initiative, led by the Transplantation Society and the International Society of Nephrology in 2008, the *Declaration of Istanbul on Organ Trafficking and Transplant Tourism* was developed during a summit attended by more than 150 representatives of scientific and medical bodies. This initiative was specifically intended to address the urgent and growing problems of organ sales, transplant tourism and trafficking. In tackling organ trafficking and transplant tourism, this international document has encouraged the development of: programs to prevent organ failure; national self-sufficiency in organ transplantation; and enhanced deceased organ donation programs.

In 2007 another initiative, the Asian Task Force on Organ Trafficking, was led by an international working group to develop a set of recommendations on how to tackle the issue of organ trafficking, particularly in Asia. For instance, it recommends health authorities to: establish a monitoring system and national registry for organ transplantation; rely more on deceased organ donation; restrict transplantation to donors and recipients from the same nationality; and address the needs of the population who suffer from economic disadvantages.

It is more than a decade that the United Nations has been fighting against human trafficking and has issued several documents regarding organ commercialism and trafficking. However, in its efforts on this issue the United Nations has recently recognized the link between organ trafficking and human trafficking. Following the UN General Assembly Resolution 63/14 of December 2008, a research plan on "organ trafficking and human trafficking" was jointly conducted by the United Nations and the Council of Europe. The idea was to establish some essential facts that would facilitate policy formulation and norm setting regarding trafficking of organs as well as trafficking in human beings for the purpose of the removal of organs.

Following the above mentioned study carried out jointly by the Council of Europe and the United Nations in 2009, the Council of Europe has recently focused on the possibility of criminalizing organ trafficking (Council of Europe 2012). It should be noted that all previous efforts have not yet been effective enough to stop the rapid growth of organ markets and trafficking.

As mentioned earlier, the above international documents have built their argument against organ trafficking and transplant tourism mostly based on exploitation, and not by addressing on discrimination and stigmatization. This oversight was recently addressed by the *UNESCO Report on the Principle of Non-discrimination and Non-stigmatization* (2014).

9.3 UNESCO's Principles of Non-discrimination and Non-stigmatization

The UNESCO International Bioethics Committee (IBC) *Report on the Principle of Non-discrimination and Non-stigmatization* (2014) has focused on the global problem of organ transplantation and trafficking in the context of the principles of Non-discrimination and Non-stigmatization as stipulated in Article 11 of the *Universal Declaration of Bioethics and Human Rights* (2005). Article 11 of this Declaration provides that, "No individual or group should be discriminated against or stigmatized on any grounds, in violation of human dignity, human rights and fundamental freedoms" (UNESCO 2005). The general aim of this principle is that in any decision or practice, no one shall be subjected to discrimination based on any grounds, including physical, mental, or social conditions, diseases or genetic characteristics, nor shall such conditions or characteristics be invoked or used to stigmatize an individual, a family, or a group (Guessous 2013). The UNESCO report also intends to ensure that organ transplantation programs further human rights and upholds the highest ethical principles especially as it pertains to Article 3 (Human dignity and human rights), Article 4 (Benefits and harm), Article 8 (Respect for human vulnerability and personal integrity), Article 10 (Equality, justice and equity), as well as Articles 11 and 13 (Solidarity and cooperation) of the Declaration.

The UNESCO Report on Non-discrimination and Non-stigmatization defines discrimination as, "From the wording of the principle one can deduce that only those distinctions that impair human dignity, human rights or fundamental freedoms are rightfully called discriminatory under Article 11. A decision or a practice that is discriminatory is one that infringes upon these fundamental notions and such decisions or practices are objectionable".

It also defines stigmatization as, "In contrast to discrimination, stigmatization is more of a social concern than a legal concept. In its more common meaning, a stigma is a mark of shame, disgrace or discredit" (UNESCO-IBC Report 2014). In this Report the issue of discrimination and stigmatization in organ transplantation and trafficking has been chosen along with five other issues as contextual examples: neurosciences; nanotechnologies; biobanking; neglected tropical diseases; and HIV/AIDS.

Article 11 is firmly rooted in the international human rights law. The most prominent of these references is the *Universal Declaration of Human Rights* of 1948 which in its first article states that: "All human beings are born free and equal in dignity and rights". Furthermore, it is important to mention that Article 7 of the Universal Declaration of Human Rights is specifically concerned with discrimination and states that: "All are equal before the law and are entitled without any discrimination to equal protection of the law. All are entitled to equal protection against any discrimination in violation of this Declaration and against any incitement to such discrimination" (UN Declaration 1948).

In order to prevent discrimination and stigmatization, the UNESCO IBC Report recommends a course of action which includes, the implementation of internationally recognized standards in organ transplantation; more restrictive rules to ensure a close and stable relationship between live donors and recipients, such as a minimum prenuptial period if a woman is a candidate for organ donation because the exploitation of women is still a serious risk in many countries; in case of organ donation by a family member, the confidentiality of information on the compatibility between the potential donor and the recipient to prevent stigma within the family if a member chooses not to donate her or his organ; effective freedom of choice of all members of the family; the prohibition of financial assistance for organ transplants abroad if the organs have been deemed objects of organ trafficking; the avoidance of discriminating against or stigmatizing victims of these violations. The Report also calls for development of an international convention against Trafficking in Human Organs (UNESCO-IBC Report 2014).

9.4 Recognizing Discrimination and Stigmatization in Organ Transplantation

The fundamental equality of all human beings has been emphasized in several international declarations and conventions such as the *Universal Declaration of Human Rights* (1948) and the *UNESCO Universal Declaration on Bioethics and Human Rights* (2005).

Accordingly, healthcare including life-saving transplantation must be allocated so that all people in equal need of a transplant have a fair opportunity to receive this treatment. Furthermore, as required by the international guidelines (WHO 2010), organs for transplantation should become available by donation free of inducement, coercion, or stigmatization. Therefore, it is necessary to take action to prevent discrimination and stigmatization in organ transplant programs as well as take action to prevent organ trade and trafficking. To this end it is important to explore how organ recipients may become subject to discrimination while in need of life saving organ transplantation, and how living organ donors may become subject to stigmatization. It is equally critical to see how organ sellers –in an organ trafficking process– become subject to stigmatization and/or discrimination.

The current international initiatives have built their arguments mostly based on exploitation inherent in organ sales and trafficking, but none of those initiatives has considered the perspectives of vulnerable patients as organ recipients and poor people as organ providers, and the discrimination and stigmatization they experience. Transplant programs that are not based on an ethically sound policy or rely on organ markets pose serious problems to justice and involve discrimination at the allocation end (in the selection of an organ recipient) as well as stigmatization of organ providers at the procurement end.

Although, the risks of discrimination and stigmatization are some of the major concerns in organ trafficking and transplant tourism, these risks exist in a legitimate organ transplant program as well. These risks become serious especially if the organ procurement and allocation programs are not based on ethical guidelines issued by international organizations such as the WHO Guiding Principles (2010) or if the ethical standard has not been fully observed and implemented.

9.4.1 The Risk of Discrimination in Organ Allocation

Allocation of scarce life-saving resources, such as human organs, requires a fair and efficient policy to avoid discrimination of patients who need organs for transplantation.

According to the international recommendations contained in the WHO Guiding Principles and Declaration of Istanbul, organs for transplantation should be equitably allocated to transplant recipients, otherwise the risk of discrimination exists. In practice, in some cases, patients who need organ transplantation become marginalized and sometimes ignored based on non-medical justifications. Examples of characteristics that should not be grounds for organ allocation or disqualifications are gender, race, religion, as well as social and financial status. Discrimination and stigmatization occur when characteristics that are not based on medical justification become the basis for organ allocation. For instance, if in public healthcare services, organ allocation is based on the "ability to pay" (as a basis for determining who gets the available organ), such a policy would discriminate against poor patients and thus is ethically unacceptable.

9.4.2 The Risk of Stigmatization in Organ Procurement

In the case of organ procurement, the risk of stigmatization is another major concern. Stigmatization may especially occur when transplantable organs become available through organ markets, transplant tourism and trafficking. A market-based organ procurement program that offers incentives to someone who has a great need is morally questionable. An irresistibly attractive offer may force poor people to make decisions to sell their organs out of desperation and against their better judgment. As Robert Veatch argues, such attractive offers become exploitative when the one making the offer has the responsibility (i.e. the government) for meeting that need (Veatch 2000). Many studies have shown that organ sellers are mostly desperate people who need money to survive or to pay their debts (Biller-Andorno and Alpinar 2013; Awaya et al. 2009). Several anthropological studies have shown how organ sellers suffer stigma in their communities (Zargooshi 2001), and even because of the fear of stigmatization they are reluctant to attend follow up donor clinics after organ removal (Ghods and Mahdavi 2007). However, as mentioned earlier, in organ procurement, discrimination may also occur. For instance the likelihood of an organ donation by a young, single, and female member of the family to a family member is greater than the likelihood of a donation by other members of the family. Furthermore, in some parts of the world, for people who provide organs, either through an altruistic donation or commercialism, there are also risks of discrimination in the work place as well as when applying for health or life insurance. In terms of employment, the one who has donated an organ may have a lower chance of getting hired or continuing her work if employers are aware of her organ donation. Based on current insurance policies, health and life insurance costs are greater for organ donors than for those who have not donated an organ.

Another form of discrimination may occur when organ procurement is based on "presumed consent". According to a presumed consent policy (opt out system), organs can be removed for transplantation after death unless individuals have objected during their lifetime (Rithalia et al. 2009). Such model aims to increase the number of transplantable organs available from deceased donors, and an individual's organs can be removed after death without their explicit consent. This policy can discriminate against people who do not or cannot express their wish not to donate due to social exclusion or because of ignorance of the legislation regarding presumed consent.

9.4.3 Transplant Tourism: Yet Another Unethical Aspect of Organ Trafficking

Organ transplants are currently performed in 91 countries with various technical advancements and regulatory oversights. According to the Global Observatory on Donation and Transplantation Report in 2010, each year 106,879 solid organs have been transplanted, which covers only 10 % of global needs (GODT 2010). The increasing gap between organ supply and demand has been cited to explain the reason behind the unethical practice of organ trafficking and illegal organ sale. As a result, patients have to travel beyond geographical borders to receive transplants because of donor shortages at home or due to unavailable transplant technology in their countries. This situation has opened the door for a market operation in human organs. Mafia-like organizations and middlemen exploit the situation, black markets are growing, and organized organ trafficking is expanding worldwide (Bagheri 2007). In transplant tourism the major sources of organs for rich patients are from underprivileged and vulnerable populations in resource-poor countries. It has been estimated that organ trafficking accounts for 5–10 % of the kidney transplants performed annually throughout the world (Budiani-Saberi and Delmonico 2008). Organ trafficking always involves organ commercialism; it ignores the dignity of organ providers and is an unethical practice to address the organ shortage and a patient's need for organ transplantation. "Organ trafficking and transplant tourism violate the principles of equity, justice and respect for human dignity. Transplant commercialism also targets impoverished and otherwise vulnerable donors; therefore, it leads inexorably to inequity and injustice" (Istanbul Declaration 2008).

Several studies have documented the long-lasting negative health, economic, psychological and social consequences of organ trafficking and transplant tourism (Cohen 2013; Naqvi 2007; Goval et al. 2002), and this negative impact along with the sense of stigmatization make it very difficult for organ providers to live in their communities. Therefore, the risks of discrimination and stigmatization are very serious when organs for transplantation come from unethical (also illegal in many jurisdictions) practices of organ trafficking and tourism, Individuals, who donate for organ trafficking or transplant tourism, are being discriminated against because of their social and/or financial status and are at risk of stigmatization in their communities after organ removal. In reality, vulnerable populations, such as illiterate and impoverished individuals, undocumented immigrants, prisoners, and political or economic refugees, in resource-poor countries are now a major source of organs for local patients through commercialism of human organs as well as for rich patient-tourists who are prepared to travel and can afford to purchase organs (Delmonico 2009). Organ traffickers use coercive means, such as force or threats of force, thus, poor people who are trafficked for the removal of their organs are in fact the victims of a crime, but unfortunately, they are also often stigmatized as criminals. In transplant tourism, individuals who are subject to trafficking for the removal of their organs, or become organ providers are trapped in a vicious cycle of exploitation, discrimination and stigmatization. They first become discriminated against as an organ provider in their community. Then they are exploited by the brokers and become stigmatized in their community because their organs have been removed. In some case they do not go back to their communities after being involved in organ trafficking.

Another issue is the link between organ trafficking and trafficking of human beings for the removal of their organs. This was first brought to international attention in 1997 by the "Bellagio Task Force Report on Transplantation, Bodily Integrity and the International Traffic in Organs" (Rothman et al. 1997). It also should be noted that in organ trafficking and transplant tourism, the concern is the exploitation and stigmatization of not just poor people as organ providers but also of the recipients of these organs through trafficking and tourism. In fact, "transplantation care for the recipients requires an expertise best provided by a transplant center that is devoted to the patient's interest, not to the patient's resources" and it is difficult to ensure that the organ has been obtained from a safe source (Delmonico 2009).

While the international focus is on cross-border organ trafficking, local organ trafficking and tourism are also causes discrimination to organ donors. As De Castro (2013) pointed out, "…local tourists appear to have escaped the regulatory radar. Authorities need to address the problem since the harm resulting from transplant tourism within national boundaries can be even more harmful and exploitative than international transplant tourism".

9.5 Conclusion

The risks of discrimination and stigmatization exist in organ transplantation as well as in organ trafficking. The UNESCO International Bioethics Committee's report on the principles of non-discrimination and non-stigmatization, based on the *UNESCO Declaration on Bioethics and Human Rights*, elaborates the possibility of discrimination and stigmatization of both organ recipients and providers, in organ transplantation as well as in organ trafficking. It calls all stakeholders to take measures to prevent discrimination and stigmatization in organ transplantation and fight against organ trafficking.

Firstly, organ procurement programs should ensure that their policies and procedures comply with the principles of non-discrimination and non-stigmatization. Next, programs should work to develop fair and transparent policies of organ allocation in which there is no discrimination of individual patient recipients. Thirdly, programs should ensure that human organs become available with no risk of stigmatization of the donors. Finally, it is important for transplant programs to actively seek ways to provide public education and awareness.

References

Awaya, T., L. Siruno, S.J. Toledano, et al. 2009. Failure of informed consent in compensated non-related kidney donation in the Philippines. *Asian Bioethics Review* 1: 2.

Bagheri, A. 2007. Asia in the spotlight of international organ trade: Time to take action. *Asian Journal of WTO and International Health Law and Policy* 2(1): 11–24.

Bagheri, A., and F. Delmonico. 2013. Global initiatives to tackle organ trafficking and transplant tourism. *Journal of Medicine, Healthcare and Philosophy* 16: 887–895.

Biller-Andorno, N., and Alpinar, Z. 2013. Organ Trafficking and Transplant Tourism. In *Handbook of global bioethics,* ed. H. ten Have and B. Gordijn. Dordrecht: Springer.

Budiani-Saberi, D.A., and F.L. Delmonico. 2008. Organ trafficking and transplant tourism: A commentary on the global realities. *American Journal of Transplantation* 8(5): 925.

Cohen, I. Glenn. 2013. Transplant tourism: The ethics and regulation of international markets for organs. *Journal of Law, Medicine & Ethics (global health and the law)* Spring: 269–285.

Council of Europe. 2012 available at: http://www.coe.int/t/DGHL/STANDARDSETTING/CDPC/PC_TO_en.asp. Last visited 25 Feb 2013.

De Castro, L. 2013. The declaration of Istanbul in the Philippines: Success with foreigners but a continuing challenge for local transplant tourism. *Journal Medicine, Healthcare and Philosophy.* doi:10.1007/s11019-013-9474-4.

Delmonico, F.L. 2009. The hazards of transplant tourism. *Clinical Journal of the American Society of Nephrology* 4: 249–250.

Ghods, A.J., and M. Mahdavi. 2007. Organ transplantation in Iran. *Saudi Journal of Kidney Disease and Transplantation* 18(4): 648–655.

Global Observatory on Donation and Transplantation. 2010. Report available at: http://www.transplant-observatory.org/Pages/home.aspx. Last visited 20 Jan 2014.

Goval, M., R.L. Mehta, L.J. Schneiderman, and A. Sehgal. 2002. Economic and health consequences of selling a kidney in India. *JAMA* 288: 1589.

Guessous, N. 2013. Non-discrimination and stigmatization. In *Handbook of global bioethics*, ed. H. ten Have and B. Gordijn. Dordrecht: Springer.

Istanbul Declaration on Organ Trafficking and Transplant Tourism. 2008. Available at: http://www.declarationofistanbul.org/. Last visited 25 Feb 2014.

Naqvi, A. 2007. A socio-economic survey of kidney vendors in Pakistan. *Transplantation International* 20: 909.

Rithalia, A., C. McDaid, S., Suekarran, et al. 2009. Impact of presumed consent for organ donation on donation rates: A systematic review. *BMJ* 338: a3162. doi:10.1136/bmj.a3162.

Rothman, D.J., E. Rose, T. Awaya, et al. 1997. The Bellagio Task Force Report on Transplantation, bodily integrity and the international traffic in organs. *Transplant Proceedings.*

The *United Nations protocol to prevent, suppress and punish trafficking in persons, especially women and children* (adopted by General Assembly resolution in 2000). Available at: http://www.uncjin.org/Documents/Conventions/dcatoc/final_documents_2/convention_%20traff_eng.pdf. Last visited 25 Feb 2014. Also see; The Universal Declaration of Human Rights 1948, available at: http://www.un.org/en/documents/udhr/index.shtml

UNESCO. 2005. Universal Declaration on Bioethics and Human Rights. Available at: http://www.unesco.org/new/en/social-and-human-sciences/themes/bioethics/bioethics-and-human-rights/. Last visited 25 Feb 2014.

UNESCO. 2014. Report on the Principle of non-discrimination and non-stigmatization. Available at: http://www.unesco.org/new/en/social-and-human-sciences/themes/bioethics/international-bioethics-committee/reports-and-advices/. Last visited April 2014.

Veatch, R.M. 2000. *Transplantation ethics*. Washington, DC: Georgetown University Press.

WHO Guiding Principles on Human Cell, Tissue and Organ Transplantation. 2010. Available at: http://www.who.int/transplantation/en/. Last visited 25 Feb 2014.

Zargooshi, J. 2001. Iranian kidney donors: Motivations and relations with recipients. *Journal of Urology* 165: 386.

Chapter 10
Dust of Wonder, Dust of Doom: A Landscape of Nanotechnology, Nanoethics, and Sustainable Development

Fabio Salamanca-Buentello and Abdallah S. Daar

Abstract Nanotechnology is a relatively recent and very promising area of inquiry devoted to the manipulation of matter at the atomic and molecular scales. Its wide reach ensures extensive influence over a vast range of human activities and has generated serious concerns over the ethical, economic, environmental, legal, and social issues (E3LSI) related to its development and applications, particularly in terms of the emergence of a "nano-divide" between high-income countries and the developing world. In this chapter, we review the advances in nanotechnology most likely to benefit low- and middle-income countries. Then, we examine the most relevant and realistic E3LS challenges related to nanotechnology (NE3LS). Next, we propose potential approaches to address these challenges, based upon foundations of equity, justice, non-discrimination, and non-stigmatization. Finally, we highlight the leading role of UNESCO in the global discussion of NE3LS issues and we suggest future pathways by means of which UNESCO's involvement in nanotechnology can contribute to the well-being of human populations worldwide.

10.1 Introduction

> "We owe it to the millions of poor people worldwide to ensure that every step we take gets us closer to a world without poverty and deprivation, and indeed, nanotechnology does have the potential to contribute towards our ability to achieve these goals in an unprecedented way. It is up to us to be bold and imaginative enough to seize this opportunity."[1]
>
> **Derek Hanekom**
> *Minister of Science and Technology of South Africa*
> *Opening speech of the World Nano-Economic Congress, 2007*

---

[1] http://www.info.gov.za/speeches/2007/07042412451001.htm

F. Salamanca-Buentello
Institute of Medical Science, University of Toronto, Toronto, Ontario, Canada

A.S. Daar (✉)
Dalla Lana School of Public Health, University of Toronto, Toronto, Ontario, Canada
e-mail: a.daar@utoronto.ca

Nanotechnology is a broad umbrella term that encompasses a wide range of relatively recent, intensely multidisciplinary, innovative research efforts involving the manipulation of matter at the atomic and molecular scale. This discipline can be defined as the study, design, creation, synthesis, manipulation, and application of functional materials, devices, and systems through control of matter at the nanometre scale (1–100 nanometres, one nanometre being equal to 1×10^{-9} of a meter), and the exploitation of novel phenomena and properties of matter that usually appear at that scale (Salamanca-Buentello et al. 2005). The convergence of a vast array of sub-disciplines and the difficulty in predicting the new horizons of nanotechnology research and development greatly complicates demarcating the scope and reach of this emergent technology (Joachim 2005; Schummer 2007).

Nanotechnology will probably have a considerable impact on many areas of human endeavour, particularly on energy storage, production, and conversion, water treatment and remediation, food and agriculture enhancement, diagnosis and treatment of disease, manufacturing, international trade, labour markets, the workplace, systems of communication, defense, international relations, civil liberties, and perhaps even the definitions of "life" and "human" (Arnall 2003; The Royal Society 2004). Such wide influence leads to concerns over the ethical, economic, environmental, legal, and social issues (E3LSI) that could theoretically result from advances in nanotechnology (Schummer 2007). A new discipline, *nanoethics*, modelled after bioethics, is struggling to emerge and still needs solid theoretical and methodological frameworks (Allhoff et al. 2007; Mnyusiwalla et al. 2003; Susanne et al. 2005). No truly novel E3LS issues seem specific to nanotechnology (NE3LSI). Challenges resulting from developments in this field have already been examined extensively in relation to previous technological waves. Keiper (2007) claims that discussions related to NE3LSI have focused too much on the hyped promises and fears of exceedingly speculative scenarios, both utopic and apocalyptic, about the hypothetical ramifications of theoretical technologies that may prove to be impossible to develop. Unrealistic assumptions underlie promises that nanotechnology will lead to "molecular manufacturing" (Drexler 1986), or the manipulation of atoms one by one, and to a posthuman cyborg-like species possessing exceptional physical and mental capabilities; equally exaggerated worries augur a catastrophe precipitated by aggressive, out-of-control, locust-like nanomachines that would wipe out all life on earth, covering the planet in a suffocating layer of "grey goo", as fancifully imagined in Michael Crichton's novel *Prey* (Arnall 2003; Baber 2004; Kulinowski 2004). To avoid becoming marginalized, nanoethicists must critically evaluate nanotechnology, collaborating with serious nanoscientists and nanotechnologists to elude unsophisticated, shallow, and unrealistic scenarios ("Don't believe the hype" 2003).

This chapter examines realistic and proximate areas of nanotechnology with the greatest risk of increasing inequality, vulnerability, discrimination, and stigmatization, with particular attention to low- and middle-income countries (LMICs). We first summarize advances in nanotechnology that could benefit the developing world and discuss the impact of nanotechnology activity in LMICs. The next section discusses the most relevant NE3LSI challenges. We then propose potential approaches to address these challenges, based upon foundations of equity, justice, non-discrimination, and non-stigmatization. Finally, we describe the contributions of UNESCO to the NE3LS discussion and we advance possible ways to enhance its essential role in the use of nanotechnology towards the solution of the most pressing global needs.

10.2 Nanotechnology for the Developing World

Science and technology are critical to achieve sustainable development. Nanotechnology offers considerable advantages over current technologies to respond to global challenges (Court et al. 2004, 2005, and 2007). Research, development, and innovation (RDI) in this area can address displacement of traditional markets, imposition of foreign values, fear that technological advances will be extraneous to development needs, and lack of resources to establish, monitor, and enforce safety regulations. We have identified the ten nanotechnology applications most relevant to the developing world and have correlated them with the Millennium Development Goals (Salamanca-Buentello et al. 2005). Based upon this and other studies, we outline below nanotechnologies with potentially beneficial influence over sustainable development.

10.2.1 Energy Production, Storage, and Conversion

Manipulation of matter at the nanoscale can provide developing countries with clean, affordable, robust, reliable, and easily maintained and serviced applications to harness renewable resources, averting recurrent energy crises, dependence on non-renewable and contaminating energy sources, and environmental degradation brought about by the depletion of oil and coal. Relevant examples of the use of nanotechnology in this area include high-efficiency solar cells, some of which could be sprayed onto any surface; ultrathin films of semiconducting polymers and nanocomposites for solar cells; quantum dot based organic light-emitting devices; nanocatalysts; carbon nanotubes for batteries and supercapacitors and, together with other lightweight nanomaterials, for robust hydrogen storage systems; nanomaterials for strong, flexible, and efficient electricity distribution; and biological-based systems for energy transduction (Mao and Chen 2007; Serrano et al. 2009).

10.2.2 Water Treatment and Remediation

Inexpensive, easily transportable, and easily cleanable water filtration nanosystems could dramatically improve water treatment and remediation. Applications that can benefit LMICs include filters based on carbon nanotubes, advanced nanomembranes, and nanoclays for water purification, detoxification, and desalination; nanoelectrocatalytic systems for decomposition of organic pollutants and removal of salts and heavy metals; magnetic nanoparticles and nanoporous materials such as zeolites and attapulgite for absorption of toxic heavy metals, organic pollutants, and micro-organisms, enabling the retrieval and recycling of contaminating substances; and nanosensors for the detection of pathogens and of inorganic contaminants (Hillie and Hlophe 2007; Qu et al. 2013).

10.2.3 Environmental Pollution Remediation

Nanotechnology-based systems can help address problems related to environmental remediation and ecosystem management. Developing countries can take advantage of titanium oxide nanoparticles and other photonanocatalysts for paints and urban coatings to deactivate and destroy air pollutants; nanodevices for the detection, absorption, and separation of toxic gases; and nano-based systems for storage and analysis of exhaustive and up-to-date massive biodiversity databases (Karn et al. 2009).

10.2.4 Prevention, Diagnosis, Monitoring, and Treatment of Disease

Advances in nanotechnology are already being used for the diagnosis and treatment of several illnesses. Nanotechnology, in tandem with genomics, has brought the promise of personalized, individualized medical diagnosis and treatment (sometimes called "theranostics") closer to reality. Quality of life in the developing world could improve through the use of microfluidic devices (labs-on-a-chip) and biosensor arrays based on carbon nanotubes, magnetic nanoparticles, quantum dots, dendrimers, nanowires, and nanobelts for inexpensive, easy to use, highly sensitive and specific, robust, portable, handheld point-of-care diagnostic kits in local clinics with the capacity to detect the presence of different pathogens (or different strains of the same pathogen) simultaneously using a minimal quantity of a single biological sample; nanoparticle systems for medical imaging; nanodevices based on nanotubes and other nanoparticles for *in situ* monitoring of monitor the concentrations of physiological variables such as glucose, carbon dioxide, and cholesterol; novel delivery systems for the slow and targeted release of drugs and for thermostable, single-dose, needle-free vaccines that increase shelf life and reduce required dosages and transportation costs (ideal for places with no adequate drug storage capabilities and distribution networks); antibody-bound nanocapsules, liposomes, dendrimers, buckyballs, nanobiomagnets, and attapulgite clays for therapeutic nanosystems that can target specific cells and tissues; and nano-based applications for regenerative medicine and medical prosthetics (Chakraborty et al. 2011; Hauck et al. 2010; Jiang et al. 2007; Martinez et al. 2010; Sosnik and Amiji 2010)

10.2.5 Agricultural Productivity Enhancement and Food Processing and Storage

Inexpensive agricultural applications of nanotechnology have the potential to decrease malnutrition, and thus childhood mortality, by increasing soil fertility and crop productivity, especially in rural regions of the developing world, while reducing the use of water, fertilizer, and pesticides, thereby also decreasing the price of agricultural products. LMICs could benefit from zeolites and other nanoporous materials that can form well-controlled stable suspensions with absorbed or adsorbed substances for the slow release and efficient dosage of fertilizers for plants and of nutrients and drugs for livestock; and from nanosensors for the detection of pathogens in livestock and plants and for crop and aquaculture monitoring. Nano-based methods of food packaging and storage may increase shelf life, enabling a wider and more efficient distribution of food products to remote areas in less industrialized countries. For example, nanobiosensors can help detect food contamination by different pathogens and antimicrobial nanoemulsions can prevent the contamination of food, equipment, and packaging, while preserving natural flavours (Chen and Yada 2011; Duncan 2011; Rai and Ingle 2012).

10.2.6 Nanotechnology Activity in LMICs

Governments worldwide have invested more than US$65 billion in nanotechnology since 2000 (http://www.cientifica.com/research/white-papers/global-nanotechnology-funding-2011/) and RDI in this field is expected to generate US$2.5 trillion a year globally by 2015 (http://www.luxresearchinc.com/blog/2010/02/the-recessions-impact-on-nanotechnology/). Financing for this field has increased exponentially in the past decade.

Several LMICs have started developing nanotechnology for their most pressing developmental challenges utilizing existing resources and capabilities, many with a view to reducing domestic inequalities and dependence on passive technology transfer from industrialized countries. Nations with a particularly active nanotechnology sector include China, India, Brazil, South Africa, Mexico, Thailand, Philippines, Sri Lanka, Vietnam, Egypt, Iran, Nigeria, Chile, Argentina, Cuba, Colombia, and Costa Rica. Developing nations have established international partnerships both with industrialized countries and among themselves. The former, while productive, tend to reproduce a pattern of technology transfer between unequal partners. Collaboration among LMICs, in contrast, can be more equitable, based on common strengths, challenges and ways to address them. For instance, Mexico and India have participated in joint projects on nanoherbicides, while Brazil, India, and South Africa have collaborated on nanomedicine, on nanoapplications for energy, water, and agriculture, and on common educational and research programs. (Court et al. 2004, 2007; MacLurcan 2012; Meridian Institute 2005; Woodrow Wilson International Center for Scholars 2007).

Nanotechnology activity in the developing world is difficult to assess because of unsettled definitions, standards, performance indicators such as number and impact of publications, number and impact of patents, number of researchers actually involved in the field, and levels of government and private sector funding. Additional obstacles include issues of categorization, language barriers, political biases, and a tendency to report what is planned instead of what has been achieved. Nevertheless, several studies (Court et al. 2004, 2007; MacLurcan 2012; Meridian Institute 2005) have examined nanotechnology engagement in LMICs, finding that most nanotechnology developments take place in the industrialized world, and that most LMICs have little or no nanotechnology activity, with considerable variability in levels of RDI funding and support. Barriers to nanotechnology development in LMICs include defective infrastructure; lack of capacity for multidisciplinary cross-sectorial collaboration; need for stable and sustained long-term science and technology activity; lack of translational lab-to-village capacity; excessive centralization of RDI; widespread corruption; inadequate government policy, including environmental and worker safety regulations; poor law enforcement; disproportionate dependence on a small number of commodities for employment, government revenue, and export earnings; deficient scientific, technical, and professional training; difficulties to establish and retain a critical mass of nanotechnology researchers; and incipient collaboration among academia, government, and industry (Court et al. 2004, 2007; Hassan 2005; MacLurcan 2012; Barker et al. 2011). It is encouraging that LMICs that are active in this field have focused on practical issues and not on hypothetical and speculative applications.

10.3 Nanotechnology Risks and Challenges

Nanotechnology is a young and rapidly developing discipline. The most pressing and realistic NE3LS concerns related to this field are its potential to both increase and decrease inequities, and the possible hazardous effects of nanomaterials on human health the environment (Malsch 2005; Roco and Bainbridge 2001; Roco 2003; Sheremeta and Daar 2004; Schummer 2007). Permeating these concerns are issues of fear and trust that need to be addressed.

10.3.1 Equity and Justice

Poverty and other social problems cannot be solved by technology alone. Addressing sustainable development challenges cannot be simply a matter of identifying technical problems and developing technological solutions to overcome them (Kenny and Sandefeur 2013). Human societies cannot be understood exclusively in flawed reductionist and mechanical terms. There is no single universal developmental path for all societies. It has been argued that science and technology are not neutral and are deeply embedded and influenced by the social context from which they arise and whose systemic inequities they can perpetuate. According to this view, novel technologies are not necessarily desirable, needed, or even inevitable, and they are not always better than previous technologies (Hillie and Hlophe 2007; Invernizzi et al. 2008; MacLurcan 2012). Conceptions of nanotechnologies as solutions to developmental issues are numerous and varied, as are entry costs, approaches to the nature and magnitude of barriers to their use, the problems that can be addressed using nanotechnology, and the infrastructure needed. Hypothetically, the very features that make nanotechnology suitable for vulnerable populations worldwide might backfire and harm them. The unwise use of or limited access to nanotechnology could precipitate a "nano-divide" between countries and individuals, exacerbating the already marked resource and power disparities between the rich and the poor further increasing the vulnerability of much of the human population to poverty, disease, inequities, exploitation, discrimination and, to a lesser extent, stigmatization (Arnall 2003; Invernizzi and Foladori 2007; MacLurcan 2012; Meridian Institute 2005). The 2014 report published by the UNESCO International Bioethics Committee provides a closer look at the potential risks of discrimination and stigmatization as a result of recent advances in nanotechnology (UNESCO 2014).

Unreal and unfulfilled expectations, unanticipated consequences, and exclusion from access to nanotechnology and its benefits could lead to resentment and social disruption. Nanotechnology could dramatically increase unequal wealth distribution, consolidating economic and social power in the private sector, particularly in multinational corporations, to the detriment of the public sector. Powerful interests could monopolize and control all aspects of nanotechnology RDI, including the design, production, and commercialization of applications and products. Apparent nano-fuelled economic growth could conceal oppression of the poor and of developing countries by the industrialized world. Market forces may drive nanoapplications and nanoproducts at the expense of developmental needs, biasing nanotechnology RDI towards the wants of the wealthy and not towards the needs of the poor (Invernizzi and Foladori 2005, 2007).

Risks could be externalized onto vulnerable populations in the absence of adequate regulations, especially if markets and commercial prospects, instead of local needs, drive nanotechnology RDI (Invernizzi et al. 2008; MacLurcan 2012). Such circumstances may stifle nano-innovation and harm fragile economies. Replacement of natural products such as export crops, minerals, and textiles, by nanotechnology-based products and materials may damage the livelihoods of the poor, decreasing demand for agricultural, mineral, and other non-fuel goods (http://www.etcgroup.org/documents/ETC_DOTFarm2004.pdf). Ninety-five developing countries derive around half of their export earnings from such commodities, but nanotechnology could make these products redundant (Barker et al. 2011).

Nanotechnology applications for agriculture and food production could decrease costs and increase crop yields using less physical, human, and financial resources, but they could also result in widespread social instability as rural workers worldwide are deprived of their livelihood (http://www.etcgroup.org/content/potential-impacts-nano-scale-technologies-commodity-markets-implications-commodity-dependent). Productivity gains may only benefit economically powerful industrial agriculture. Industrial production of nanotechnologies and nanomaterials could exhaust critical material resources, weakening labour and generating waste (Schummer 2007; Scrinis and Lyons 2007). According to the United Nations Conference on Trade and Development, two billion individuals are employed in commodity production. Advanced nanomaterials could substitute for rubber, carbon nanotubes could replace copper wires, and platinum could be substituted by nano-alloys, devastating the economies of countries such as Chile or Zambia that depend on their metal mining sectors. Nanoengineered polymers could replace cotton and other natural fibers, affecting LMICs that rely heavily on textile exports such as Mali (Barker et al. 2011). Novel developments in nanotechnology could lead to an increase in the exploitation of unskilled individuals for cheap labour, to pronounced job loss, and to increased migration from nano-poor to nano-rich regions.

The lack of advanced education and training in nanotechnology-related fields may lead to deepening of economic and social inequities due to the decreased capacity for competition and innovation. In particular, LMICs risk a "brain drain" of nanotechnology experts educated and trained at great expense in the developing world only to end living and working in the industrialized world. Furthermore, researchers in this field whose native language is not English typically face considerable challenges to contribute to scientific knowledge and to be taken into account in decision-making processes. Finally, nanotechnology could exacerbate the gender disparities already evident in the shortage of women in mathematics, engineering, and the physical sciences.

10.3.2 Environmental Nanotoxicity

Concern about the environmental toxicity of nanomaterials has grown in the last decade (Arnall 2003; Hett 2004). A new discipline, nanotoxicology, aims to determine whether and to what extent the novel properties of nanomaterials, especially those used industrially and commercially, affect both the environment and the human body (Oberdörster et al. 2005; Maynard et al. 2011). Terms such as "nanopollutants" and "nanowaste" are becoming increasingly used. Risks related to the environmental and health toxicity of nanomaterials are realistic and relatively straightforward to address.

Matter at the nanoscale tends to exhibit unique properties due to features such as quantum size effects, large surface area to volume ratio, shape, surface charge, and aggregation and solubility characteristics. These attributes may lead to unusual toxic effects that are considerably different from those seen at larger scales (Maynard et al. 2011). For example, it is well known that gold, inert at the macroscale, is highly reactive at the nanoscale. These characteristics of nanomaterials also complicate their removal from air, water, and soil.

Many nanomaterials, especially if non-degradable or slowly degradable, may pose a threat to the environment and to living organisms; however, the specific toxic effects and processes are poorly understood (Schummer 2007). Some nanomaterials can bioaccumulate in edible organisms and can thus become incorporated into food chains. Several studies have shown that nanomaterials such as fullerenes can cause cell and tissue damage in different species, including humans (Maynard et al. 2011; Hubbs et al. 2013). Experimental studies in animals have shown that inhaled nanoparticles can reach the brain through the olfactory nerves, that water-soluble fullerenes can generate oxidative damage in central nervous system lipids, and that nanotubes can induce inflammatory lesions in lungs.

There is a lack of proper evaluations of the complete life cycles of nanoengineered materials, including their fabrication, storage and distribution; their application and potential abuse; and their disposal, destruction and recycling. A particular concern for LMICs is the possibility that the developing world could be the dumping ground of unwanted, low-quality, or potentially toxic nanoproducts from industrialized nations.

10.3.3 Health Issues

Advances in nanotechnology may widen the gap between the cutting-edge diagnostic capabilities and the availability of therapeutic measures (Gordijn 2007). Nanodevices may replace health workers (Schummer 2007). Moreover, research on the behaviour of nanomaterials inside the human body is still in its infancy. The characteristics that make nanomaterials useful in health-related applications can potentially lead to dangerous and toxic physiological effects (Gordijn 2007; Maynard et al. 2011; Chou and Chan 2012). Most known nanomaterials are easily absorbed by inhalation, ingestion, and contact with skin and mucous membranes; they also distribute widely throughout the organism (Arnall 2003). Some nanomaterials cause inflammation, weaken the immune system, bioaccumulate in vital organs, interfere with homeostasis, and are toxic to human tissue and cell cultures. Fullerenes, metal oxide nanoparticles, and other nanomaterials with high chemical and biological reactivity can increase production of reactive oxygen species, in particular of free radicals, which generate oxidative stress, inflammation, considerable damage to cellular structures like mitochondria and cell nuclei, DNA mutations, and cell death (Chou and Chan 2012; Hubbs et al. 2013). Few precise, standardized, and sensitive quantitative and qualitative risk assessment methods exist. Reliable information on the exposure hazard of populations at risk to potentially toxic nanomaterials is scant, particularly that related to the workplace (Kuempel et al. 2012; Schulte and Salamanca-Buentello 2006). Workers exposed to nanomaterials may lack specially designed engineering controls and personal protective equipment.

10.3.4 Policy, Legal, and Intellectual Property Issues

Existing legislation, particularly in the developing world, may prove inadequate and too restrictive to address the rapidly evolving nature of nanotechnology, but an overreaction to regulatory deficiencies may lead to a heavy-handed response that may inhibit potentially beneficial RDI in the field (Hodge et al. 2010). For example, while lax regulations related to the potential toxicity of nanomaterials could encourage dumping nanowaste in the developing world, inordinately restrictive laws spark international conflict over production and transportation of nanomaterials.

Intense investment worldwide in nanotechnology has generated a massive surge in related patents filed by academia and the private sector, but aggressive patenting of nano-derived products, particularly at such an incipient stage of development of the sector, may stifle innovation and drive up costs, reducing the potential for creating and commercializing applications that could benefit low income populations in both the industrialized and the developing worlds, thus increasing inequities. A cutthroat, fiercely competitive international intellectual property system could further concentrate the ownership of nanotechnology applications and products in high-income countries (http://www.etcgroup.org/content/special-report-nanotechs-second-nature-patents-implications-global-south). Most patents and patent applications related to nanotechnology originate in high-income countries and are concentrated in a few universities and multinational corporations. About 90 % of the total patent share in health-related nanoproducts and nanoapplications is held by less than 10 countries. The vast majority of these patents is held by the private sector and by companies, not individuals (MacLurcan 2012). Nano-innovation could be severely inhibited by broad patents that cover the fundamental concepts and building blocks of nanotechnology (fullerenes, nanotubes, nanoparticles, quantum dots), along with any related processes and applications, exclusive to a single person or entity (Pearce 2012; Schummer 2007). Overreaching patents controlled by a few entities could lead to "patent thickets", dense morasses of overlapping sets of patents rights affecting wide areas of nanotechnology, thereby increasing costs, restricting technical development, and limiting access to fundamental knowledge (Sabety 2004).

Trade barriers and broad and restrictive patents over nanotechnology could have a very negative effect on the capacity of developing countries to harness nanotechnology. Most existing patents related to the use of nanotechnology in health care focus on medical conditions common in the industrialized world at the expense of neglected diseases prevalent in LMICs. Without domestic nanotechnology RDI, LMICs could have to pay exorbitant fees for the use of nanotechnologies created and patented in the industrialized world. LMICs could be further hampered by the shortage of lawyers, patent officers, policy experts, and other decision-makers with solid and up-to-date knowledge about nanotechnology.

10.3.5 Individual Autonomy and Dignity

The definitions and limits of *identity*, *normalcy, disability, health*, and *disease*, and the demarcation of what is and is not part of the natural human body, may need to be revised in light of the convergence of nanotechnology, biotechnology, genomics and the biomedical sciences, information technology, and the cognitive sciences, abbreviated as NBIC (Evans 2007; Roco and Bainbridge 2003). The reductionist view of the body as just a machine made of commercializable and replaceable parts may become more pronounced (Gordijn 2007). Nanodevices may lack well-defined parameters of human control and autonomous decision-making, complicating assignation of responsibilities and increasing the sense of lack of control over one's own body (Gordijn 2007; Schummer 2007).

Much of the initial discussion on the potential impact of nanotechnology on individual autonomy and dignity was squandered on speculations about hypothetical distant utopias (or dystopias) populated by physically, emotionally, and cognitively "enhanced" cyborg-like post-humans whose behaviour would have been made "desirable" (Arnall 2003; Evans 2007; Gordijn 2007). In contrast, the more immediate and realistic concerns relate to the discrimination and stigmatization of individuals "enhanced" through these technologies (UNESCO 2014). Paradoxically, in an ultra-competitive society, the pressure to enhance may increase; individuals who may not have access to nanobiomedical advances or who may not want to take advantage of such technologies, may also be stigmatized, discriminated against, and even, very hypothetically, "enhanced" using coercive means. Persons with disabilities would be most vulnerable to these latter risks.

10.3.6 Privacy and Confidentiality

The increasing miniaturization and effectiveness of inexpensive surveillance devices may lead to broader and highly intrusive methods of gathering data, making it widely available, and facilitating people control by powerful individuals, governments, and corporations, with the potential to severely erode people's privacy, confidentiality and human rights generally. This "nanopanopticism" could critically endanger civil liberties (Gordijn 2007; MacDonald 2004; Monahan and Wall 2007). Additionally, nanotechnology applications initially developed for agricultural enhancement could be applied towards biowarfare and population surveillance.

10.3.7 Defense and Security

Militaries are eager to harness nanotechnology to improve their offensive and defensive capabilities (Arnall 2003). Most interest focuses on soldier protection and survivability by decreasing the weight that soldiers carry, improving blast and ballistic protection, creating new methods of detecting and detoxifying chemical and biological threats, and providing physiological monitoring and automated medical intervention. Research has also been done in the creation of powerful and destructive weapons. A risk of the use of nanotechnology for military purposes is that the result of public funded RDI projects could be end up being controlled by private interests and starting a nano arms race. Furthermore, hypothetically, soldiers who refuse to be nanotechnologically "enhanced" could be segregated, dismissed, or otherwise harmed, and their rights could be restricted. However, as with discussions related to the rest of nanotechnology, analysis of the potential impact of advances in this field have suffered from simplistic political and social assumptions underlying the description of problems and the proposed solutions.

10.4 Perspectives and Opportunity

10.4.1 Increasing Equity

If the most severe contemporary global ethical issues are the major disparity between the standard of living in industrialized and developing nations and the socio-economic inequities within countries (Benatar et al. 2005), then the global community has the responsibility to judiciously harness promising tools such as nanotechnology to address the priorities of vulnerable populations, especially in the developing world, while simultaneously preventing a nano-divide.

Using nanotechnology to address development problems does not exclude acknowledging the complex sources, contexts, and dynamics behind socio-economic inequities (Parr 2005). While nanotechnology is clearly not a silver bullet, the developing world should actively pursue its use to solve its most pressing challenges in tandem with all other available strategies, from the simplest to the most complex (Singer et al. 2005) instead of passively waiting from the sidelines for solutions to arrive from the industrialized world through technology transfer (Court et al. 2007; Hassan 2005). Criteria to define and prioritize which nanotechnologies can best address developmental issues can include potential impact, burden, appropriateness, feasibility, existing knowledge gaps, and indirect benefits (Salamanca-Buentello et al. 2005). Thus, it is essential to identify priorities, resources, capabilities, limitations, potential niche areas, and opportunities for strategic engagement; nanotechnology road-mapping exercises would also be particularly useful (Mehta 2004; Thorsteinsdóttir et al. 2004).

LMICs must develop their science and technology RDI capacity, and design, develop, produce, and market their own simple, affordable, and accessible nanotechnology products and applications. Factors that may promote the successful development of nanotechnology include strong collaboration and linkages between academia, government, and industry; clear definition of niche areas, with special attention to existing natural conditions, resources, and infrastructure; a focus on particular stages of nanotechnology RDI where these countries have strengths; leveraging of competitive advantages; judicious private sector involvement; strong leadership; and political will that translates into long-term government support and funding (Thorsteinsdóttir et al. 2004). Strong national institutions that design, carry out, fund, coordinate, and supervise country-wise, multi-sector, interdisciplinary programs must be balanced with the imperative to decentralize nanotechnology RDI. Capital-intensive large scale projects that promote excessive centralization of resources (white elephants) should be avoided. Nanotechnology-centred clusters throughout each LMIC can improve efficiency by encouraging resource sharing, cross-sectoral integration, technology transfer, and the generation of a critical mass of nano experts.

"Brain drain" must be turned into "brain recirculation". To create capacity and cultivate a core of nanotechnology researchers, the developing world should take advantage of the scientific, technological, networking, management, and investment capabilities of their diasporas, the communities of individuals from a specific developing country who left home to attend school or find a better job and who now work in industrialized nations in academia, research, or industry (Séguin et al. 2006). LMICs should develop and guide their own paths based upon their most critical needs with a long-term view, without excluding the possibility of learning from the successes and mistakes of other nations. Using an analogy from the interactions between species in nature, the relationship between industrialized countries and LMICs should be symbiotic, mutualist, or at least neutral, but neither parasitizing nor predatory.

10.4.2 Assessing and Managing Risks

The most troubling nanotechnology risks may be the "unknown unknowns", the unanticipated hazards and wrong assumptions related to this field (Maynard 2006; Schulte and Salamanca-Buentello 2006). A debate exists between those who argue that nanotechnology should develop freely unless solid proof exists of concrete risks, and those who claim that all nanomaterials, nanoapplications, or nanoproducts must be shown to be completely risk-free before exposing the general public to them. We cannot assume, as some have suggested (www.etcgroup.org/sites/www.etcgroup.org/files/thebigdown.pdf), that all nanotechnologies are unsafe until proved otherwise, a view that would lead to a mandatory international moratorium on all stages of nano development. Such radical measures are neither realistic nor beneficial, as they lack empirical foundations, and would likely have more negative consequences for LMICs than for wealthy, industrialized nations. Potential benefits should not mask potential risks, but the latter should not derail conceivable advantages.

The evaluation of nanotoxicological risks should include strong international cooperation and sensitive protocols for all stages of the life-cycle of nanoproducts and nanoapplications to detect, assess, predict, control, and mitigate their most damaging effects (Burleson et al. 2004). Given the difficulties in making generalizations about the environmental and health risks of nanotechnology products, each specific nanomaterial must undergo a thorough toxicological and pharmacological study, a complete *in vivo* risk assessment, and a careful cost-to-benefit analysis (Burleson et al. 2004). Similarly, clinical approval of medically-relevant nanomaterials must be made on a case-by-case basis (Evans 2007; Gordijn 2007). Environmentally sound processes are needed to extract raw materials. LMICs should never be used as testing or dumping grounds for potentially toxic nanoproducts and nanowaste from wealthy nations.

10.4.3 Designing Adequate and Flexible Laws and Regulations

Existing laws and regulations must be modified to address the most realistic issues related to nanotechnology, particularly the concrete challenges related to discrimination and stigmatization of vulnerable populations (UNESCO 2014). New regulatory frameworks must be logical, scientifically-based, efficient, flexible, and transparent; they should be able to keep the pace of technological advances (or even anticipate them), while remaining adaptable to specific local contexts and balancing potential risks and benefits (Hodge et al. 2010; Hodge et al. 2014; Reynolds 2003). Legislation should not be over-encompassing and instead focus on specific nanomaterials, uses, and fields. Standard operational definitions are needed for all areas of nanotechnology, in particular for nanotoxicology. Similarly, it is urgent to design unified international standards for measurement of the concentration and toxicity of nanomaterials in the environment and in the human body, in particular the exposure of workers and of consumers to novel, potentially toxic nanomaterials. A global authority could create and enforce environmental standards for individuals at risk to exposure to novel nanomaterials. In this way, the international community would not only share the potential toxic hazards of nanomaterials, but also the ways to prevent them and deal with their consequences. It is also urgent to determine whether or not labelling for nano-products is desirable.

Incentives and rewards can encourage nanotechnology RDI, particularly in academic and state-controlled institutions in LMICs, while still ensuring access and availability of inexpensive nanoproducts and nanoapplications critical for developmental needs. Balancing these contrasting issues could be achieved using strategies that have proved successful in information technology such as patent pools, patent clearinghouses, and open-source approaches (Court *et al.* 2007; Lemley 2005). Patent thickets should be avoided. Mechanisms are needed to encourage individuals, institutions, and governments in LMICs to patent locally developed, socially relevant nanotechnology applications, bearing in mind that the inappropriate use of patents might inhibit RDI efforts. In particular, care should be taken to guarantee universal access to potentially life-saving nanotechnology-based medications. Pearce (2012) has suggested making the results of all publicly funded nanotechnology research available to everyone for free. The goal would be to reduce costs, enable the use of the best materials, processes, and applications available, stimulate innovation, and lower barriers for entry. Also, national and international legislation preventing the patenting of the basic building blocks of nanotechnology could be enacted, following the example of the ruling by the Supreme Court of the United States regarding the non-patentability of natural human genes (http://www.supremecourt.gov/opinions/12pdf/12-398_1b7d.pdf). Patent examiners, especially in the developing world, need access to up-to-date nanotechnology literature and must be able to handle the conceptual complexity and fast-changing technical language of this field.

10.4.4 Improving Education and Public Engagement

The sophistication of nanotechnology demands a broad and rigorous interdisciplinary education at all levels of education that prepares individuals for knowledge-based economies (UNESCO 2005; United Nations Millennium Project 2005). Such training must include a strong grounding in several fields of mathematics, science, technology, and engineering, with emphasis in computer and information sciences, given the massive amounts of data generated by most projects in the field. It should be able to adapt to rapidly changing environments and should harness online resources to encourage broad access to up-to-date nanotechnology-related information (Pearce 2012), focusing on bridging language barriers resulting from highly-specialized terminology and from the use of English as an academic *lingua franca*. While manual labour will inevitably decrease as a result of the automation boost generated by nanotechnology, other types of jobs demanding advanced skills and abilities will be created. At the same time, dependence on specialists could be reduced by designing nanotechnology-derived products and applications that could be locally maintained with ease and at low cost.

The general public needs a common knowledge base to be able to participate in decision-making related to nanotechnology. Education in this field can happen in both formal and informal contexts. The Centre for Nanosciences and Nanotechnology in Mexico has carried out a remarkable effort widely distributing a book on nanotechnology designed for the general public which is written in Spanish and has been translated into several indigenous languages. Public debate about NE3LS issues should unfold in parallel to the development of nanotechnology and should include topics such as access to reliable information, control, introduction into society, potential benefits and risks, and responsibility over negative effects on particular populations (Court et al. 2007). To avoid speculation and hype, NE3LSI discussions should focus on realistic, concrete, scientifically-based developments.

To adapt advances in nanotechnology to the particular contexts in which they will be applied, ongoing public discourse on the potential benefits and risks is essential, thereby taking into account the views of all relevant stakeholders to map the future of nanotechnology and incorporating the general population into formal decision-making processes related to nanotechnology policy (Einsiedel and Goldenberg 2004; Lewenstein 2005). Active, imaginative, resourceful, and open-minded dialogue among a broad segment of the public should be encouraged, taking care to avoid top-down, manipulative, and exclusively didactic campaigns that elicit passive acquiescence and approval, or vehement campaigns in favour (nano-marketing) or against the use of nanotechnology (http://www.etcgroup.org/fr/node/51). Public engagement must acknowledge the legitimate expectations, fears, and concerns of specific populations (Berndtson et al. 2007; Tindana et al. 2007) and must identify and understand the sources and nature of popular representations of nanotechnology (Hornig-Priest 2005). Each member of the public should be seen as an active stakeholder who can participate in broad, multi-way, candid deliberation and dialogue that enable people to exchange views fairly, objectively and respectfully (Court et al. 2007; MacLurcan 2012).

Perceptions about nanotechnology may diverge from reality, but they still shape the public's reaction to developments in this field. Attitudes may evolve or deteriorate as advanced technologies, including biotechnology and genomics, converge towards the nanoscale, and as nanotechnology applications become part of popular culture. The general public, once aware of the nature and potential impact of this technology, may wish to seek information about benefits and risks (Scheufele and Lewenstein 2005). Changing perceptions is very difficult and does not depend on simply presenting facts, although the availability of scientific-based information can reduce irrational reactions. Well-known psychological tendencies and fears increase credulity and confusion and may lead to negative perceptions that can aggravate in the absence of trust (Macoubrie 2006; Williams-Jones 2004). Decision-making has a very strong emotional component. Diverse social and cultural factors such as gender, ethnicity, social and economic standing, religion, traditions, and the framing of information colour the approval or rejection of nanotechnology (Choi 2007; Lee et al. 2005; Toumey 2012). Public engagement strategies must help people become aware of their own background, belief systems, and biases, and facilitate acknowledging and engaging with other points of view (Choi 2007). Transparency, openness, and humility are essential in the discussions of benefits and risks if public engagement strategies are to be successful.

The nanotechnology revolution may be the first scientific and technological wave in which experts are not exclusively in charge of RDI in the field. Such experts should ideally be skilled at communicating with the general public, and they should only make parsimonious claims to prevent creating unwarranted expectations or fears. The participation of specialists from the social sciences and the humanities in dialogues on nanotechnology is welcome and desirable (Macnaghten et al. 2005). But it is necessary to realize that experts, even those from the developing world, belong to a privileged, educated, and globalized academic, social, economic, and political elite and thus their views and interests may not be representative of those of the general population. Moreover, along with entrepreneurs, government officials, and other members of the ruling classes, they may favour approaches to public engagement that let them direct the terse, predetermined development of nanotechnology. Journalists and NGOs can help shape public perception and mediate between experts and the public (Choi 2007).

10.4.5 *Focusing on Nanogovernance and Nanodiplomacy*

We have put forth the concept of *nanodiplomacy* to encompass the skillful collaboration between nations to take advantage of the opportunities offered by nanotechnology to address developmental needs (Singer et al. 2006). Several strategies can be employed to achieve this goal. A first step consists of designing a global roadmap that clearly shows existing capacities and limitations in nanotechnology RDI within each country and in different world regions in order to set priorities and find synergies. Then, using solid up-to-date information, a global governance model for the responsible use of nanotechnology with a long-term view can be created (Kearnes et al. 2006). Such a model would include the development and implementation of rational and efficient regulatory regimes for risk management and benefit sharing (http://www.etcgroup.org/fr/node/51). All countries need to leverage their national nanotechnology assets into foreign policy. Cooperation between industrialized and developing countries and among LMICs should be encouraged to share resources, facilities, knowledge, experiences, and applications regarding nanotechnology RDI (Denis 2005). Such collaboration can be achieved through the creation of international networks of excellence (MacLurcan 2012). If national and global conferences and meetings are encouraged, they should always include as a core component discussions on NE3LS issues.

Public and private funding is essential to stimulate nanotechnology RDI to address sustainable development challenges. In both industrialized and developing countries, governments should provide incentives for the private sector to fund the development of nanotechnology in LMICs, especially in those nations in which academia, companies, and industries are disconnected and where technology is imported, not developed.

10.4.6 Aligning Nanotechnology with the Sustainable Development Goals

In 2000, all 189 members of the United Nations committed to achieve the eight Millennium Development Goals (MDGs) by 2015. The MDGs have been used to measure progress in human development, to assess levels of social and economic sustainability (http://www.un.org/millenniumgoals/), and to encourage actions locally, regionally, and globally that have improved the quality of life worldwide. We have correlated the MDGs with the ten applications of nanotechnology most likely to benefit the developing world (Salamanca-Buentello et al. 2005). In 2012 the world's governments agreed at the Rio + 20 Summit to adopt a new set of objectives focused on poverty alleviation, social inclusion, environmental sustainability, and good governance: the Sustainable Development Goals (SDGs) for the 2015–2030 period (Kenny 2013; Sustainable Development Solutions Network 2013; United Nations High-Level Panel of Eminent Persons 2013). Jeffrey Sachs, Director of the Sustainable Development Solutions Network, has argued that two critical tools to achieve the SDGs are nanotechnology and sophisticated, real-time data gathering and analysis (http://www.scidev.net/global/sustainability/feature/-jeffrey-sachs-world-post-mdg-future.html).

Building on the Grand Challenges in Global Health initiative (Singer et al. 2007; Varmus et al. 2003), we proposed in 2005 issuing a series of Grand Challenges in

Nanotechnology (Court et al. 2007; Salamanca-Buentello and Persad 2005; Singer et al. 2006). A grand challenge is call to arms for researchers and experts to identify specific scientific or technological barriers hindering progress towards an area of development so that appropriate and equally specific solutions can be designed and implemented to overcome such obstacles. The SDGs could guide the selection of the Grand Challenges in Nanotechnology. To ensure that nanotechnology and other advanced emergent technologies can help achieve the SDGs, universities, research centres, technical institutions, national governments, the private sector, civil society, and international institutions such as UNESCO and other United Nations agencies must be integrated into efficient and flexible networks that can provide novel and practical solutions. Developing countries must be adequately and actively represented in such networks.

10.5 The Role of UNESCO in Nanoethics

UNESCO is uniquely positioned to incorporate NE3LS elements in the development of nanotechnology. In 2005, it created an *ad hoc* committee of experts, the discussions of which resulted in a comprehensive book *Nanotechnologies, ethics, and politics* (Ten Have 2007a). UNESCO had already prepared a high-impact report on the ethical dimensions of nanotechnology, *The ethics and politics of nanotechnology*, published in its six official languages (UNESCO 2006). Furthermore, in its recent report, the International Bioethics Committee (IBC) has highlighted the potential effects of advances in nanotechnology on discrimination and stigmatization, along with possible courses of action to address these topics (UNESCO 2014).

Building upon these efforts and on its ethical mandate, UNESCO can have a decisive role in several relevant fields (Ten Have 2007b). First, based upon the Universal Declaration on Bioethics and Human Rights, it should provide the common ethical foundations for the analysis of NE3LS issues, balancing the need for universal principles with their application in specific socio-cultural contexts. UNESCO should actively participate in the identification and prioritization of these issues, ensuring that NE3LS debate does not become biased towards the views and needs of wealthy countries. Through the International Bioethics Committee (IBC) and the World Commission on the Ethics of Scientific Knowledge and Technology (COMEST), UNESCO should guide the assessment of the development and use of nanotechnologies, performing an ethical watch function and anticipating NE3LS issues related to this rapidly evolving field. Such a task can be promoted by publishing and distributing educational materials.

UNESCO should also help develop novel guidelines, standards, benchmarks, laws, and regulations related to nanotechnology and its applications, and should mediate their implementation. As an advisory body, it should participate in the design of recommendations for decision makers on policy issues related to nanotechnology. In terms of increasing awareness of NE3LS issues, UNESCO can foster education on nanotechnology and nanoethics globally at all educational levels, with emphasis on the ethical training of nanoscientists and nanotechnologists. It should also advocate for the inclusion of NE3LS components in nanotechnology-related research projects, including academic theses. The global support and perspective of UNESCO for ethics education in NE3LS will enhance the ability not only of nanotechnology experts but also of the general public to identify and critically examine challenges and to develop solutions through thoughtful decision-making. Moreover, UNESCO should contribute its expertise to build and sustain NE3LS capacities in LMICs, partly helping create, enhance, and sustain national ethics infrastructures, including ethics committees and review boards.

Taking advantage of the Global Ethics Observatory, UNESCO can become an international information exchange of NE3LS experts, centres, institutions, and organizations from the local to the global levels, teaching programs, legislation, guidelines, and policies (Ten Have and Ang 2007). Similarly, UNESCO should help organize multidisciplinary regional and global networks of experts and international *fora* that encourage both cooperation and wide public dialogue in NE3LS issues with the participation of all relevant stakeholders.

UNESCO's Expert Group on Nanotechnology and Ethics should carefully examine the potential benefits and risks associated with nanotechnology and vigorously advocate for its use to address developmental needs. The recently created UNESCO international scientific advisory board (http://www.scidev.net/global/policy/news/unesco-to-set-up-un-science-advisory-board.html) can ensure that nanotechnology is used judiciously for the well-being of vulnerable populations worldwide.

References

Allhoff, F., P. Lin, J. Moor, and J. Weckert (eds.). 2007. *Nanoethics: The ethical and social implications of nanotechnology*. Hoboken: Wiley.
Arnall, A.H. 2003. *Future technologies, today's choices: Nanotechnology, artificial intelligence and robotics*. London: Greenpeace Environmental Trust.
Baber, Z. 2004. "An undifferentiated mass of gray goo?" Nanotechnology and society. *Bulletin of Science, Technology & Society* 24: 10–12.
Barker, T.F., L. Fatehi, M.T. Lesnick, et al. 2011. Nanotechnology and the poor: Opportunities and risks for developing countries. In *Nanotechnology and the challenges of equity, equality and development*, Yearbook of nanotechnology in society, vol. 2, ed. S.E. Cozzens and J. Wetmore, 277–290. Dordrecht: Springer.
Benatar, S.R., A.S. Daar, and P.A. Singer. 2005. Global health challenges: The need for an expanded discourse on bioethics. *PLoS Medicine* 2: e143.
Berndtson, K., T. Daid, C.S. Tracy, et al. 2007. Grand challenges in global health: Ethical, social, and cultural issues based on key informant perspectives. *PLoS Medicine* 4: e268.
Burleson, D.S., M.D. Driessen, and R.L. Penn. 2004. On the characterization of environmental nano particles. *Journal of Environmental Science and Health, Para A. Toxic/Hazardous Substances and Environmental Engineering* 39: 2707–53.
Chakraborty, M., S. Jain, and V. Rani. 2011. Nanotechnology: Emerging tool for diagnostics and therapeutics. *Applied Biochemistry and Biotechnology* 165: 1178–1187.
Chen, H., and R. Yada. 2011. Nanotechnologies in agriculture: New tools for sustainable development. *Trends in Food Science and Technology* 22: 585–594.

Choi, K. 2007. Public Engagement and Education for ethics in nanotechnology. In *Nanotechnologies, ethics, and Politics*, ed. H. Ten Have, 181–204. Paris: UNESCO.
Chou, L.Y.T., and W.C.W. Chan. 2012. Nanotoxicology: No signs of illness. *Nature Nanotechnology* 7: 416–417.
Court E.B., Daar A.S., Martin E., et al. 2004. Will Prince Charles et al diminish the opportunities of developing countries in nanotechnology? *Nanotechnology* 4: 3. Available at http://www.nanotechweb.org/articles/society/3/1/1/1
Court, E.B., A.S. Daar, D.L. Persad, and F. Salamanca-Buentello. 2005. Tiny technologies for the global good. *Nano Today* 8: 14–15.
Court, E.B., P.A. Singer, F. Salamanca-Buentello, and A.S. Daar. 2007. Nanotechnology and the developing world. In *Nanotechnologies, ethics, and politics*, ed. H. Ten Have, 155–180. Paris: UNESCO.
Denis M. (ed.). 2005. Funding and support for international nanotechnology collaborations. Nanoforum.org Report. European Nanotechnology Gateway
Don't believe the hype. Editorial. 2003. *Nature* 424: 237.
Drexler, K.E. 1986. *Engines of creation: The coming era of nanotechnology*. New York: Anchor.
Duncan, T.V. 2011. Applications of nanotechnology in food packaging and food safety: Barrier materials, antimicrobials and sensors. *Journal of Colloid and Interface Science* 363: 1–24.
Einsiedel, E.F., and L. Goldenberg. 2004. Dwarfing the social? Nanotechnology lessons from the biotechnology front. *Bulletin of Science, Technology & Society* 24: 28–33.
Evans, D. 2007. Ethics, nanotechnology and health. In *Nanotechnologies, ethics, and politics*, ed. H. Ten Have, 155–180. Paris: UNESCO.
Gordijn, B. 2007. Ethical issues in nanomedicine. In *Nanotechnologies, ethics, and politics*, ed. H. Ten Have, 155–180. Paris: UNESCO.
Hassan, M.H.A. 2005. Nanotechnology: Small things and big changes in the developing world. *Science* 309: 65–66.
Hauck, T.S., S. Giri, Y. Gao, and W.C.W. Chan. 2010. Nanotechnology diagnostics for infectious diseases prevalent in developing countries. *Advanced Drug Delivery Reviews* 62: 438–448.
Hett, A. 2004. *Nanotechnology: Small matter, many unknowns*. Zurich: Swiss Reinsurance Company.
Hillie, T., and M. Hlophe. 2007. Nanotechnology and the challenge of clean water. *Nature Nanotechnology* 2: 663–664.
Hodge, G.A., D.M. Bowman, and A.D. Maynard (eds.). 2010. *International handbook on regulating nanotechnologies*. Cheltenham: Edward Elgar Publishing Limited.
Hodge, G.A., A.D. Maynard, and D.M. Bowman. 2014. Nanotechnology: Rhetoric, risk and regulation. *Science and Public Policy* 41: 1–14.
Hornig-Priest, S. 2005. Room at the bottom of Pandora's Box: Peril and promise in communicating nanotechnology. *Science Communication* 27: 292–299.
Hubbs, A.F., et al. 2013. Nanotechnology: Toxicologic pathology. *Toxicologic Pathology* 41: 395–409.
Invernizzi, N., and G. Foladori. 2005. Nanotechnology and the developing world: Will nanotechnology overcome poverty or widen disparities? *Nanotechnology Law and Business Journal* 2: 101–110.
Invernizzi, N., and G. Foladori. 2007. Nanotechnology for developing countries. Asking the wrong question. In *Assessing societal implications of converging technological development*, ed. G. Banse, A. Grunwald, I. Hronszky, and G. Nelson, 229–230. Berlin: Sigma.
Invernizzi, N., G. Foladori, and D. MacLurcan. 2008. Nanotechnology's controversial role for the south. *Science, Technology and Society* 13: 123–148.
Jiang, W., B.Y. Kim, and J.T. Rutka. 2007. Advances and challenges of nanotechnology-based drug delivery systems. *Expert Opinion on Drug Delivery* 4: 621–633.
Joachim, C. 2005. To be nano or not to be nano? *Nature Materials* 4: 107–109.
Karn, B., T. Kuiken, and M. Otto. 2009. Nanotechnology and in situ remediation: A review of the benefits and potential risks. *Environmental Health Perspectives* 117: 1813–1831.

Kearnes, M., P. Macnaghten, and J. Wilsdon. 2006. *Governing at the nanoscale: People, policies and emerging technologies*. London: Demos.

Keiper, A. 2007. (Spring). Nanoethics as a discipline? *The New Atlantis* 16: 55–67.

Kenny, C. 2013. What should follow the millennium development goals? *British Medical Journal* 346: f1193.

Kenny C., and Sandefeur J. 2013. Can silicon valley save the world? *Foreign Policy*: 72–7

Kuempel, E.D., C.L. Geraci, and P.A. Schulte. 2012. Risk assessment and risk management of nanomaterials in the workplace: Translating research to practice. *Annals of Occupational Hygiene* 56: 491–505.

Kulinowski, K. 2004. Nanotechnology: From "Wow" to "Yuck"? *Bulletin of Science, Technology & Society* 24: 13–20.

Lee, C.J., D.Z. Scheufele, and B.V. Lewenstein. 2005. Public attitudes toward emerging technologies: Examining the interactive effects of cognitions and affect on public attitudes toward nanotechnology. *Science Communication* 27: 240–267.

Lemley, M.A. 2005. Patenting nanotechnology. *Stanford Law Review* 58: 601–630.

Lewenstein, B. 2005. Nanotechnology and the public. *Science Communication* 27: 169–174.

MacDonald, C. 2004. Nanotechnology, privacy and shifting social conventions. *Health Law Review* 12: 37–40.

MacLurcan, D.C. 2012. *Nanotechnology and global equality*. Singapore: Pan Stanford.

Macnaghten, P., M.B. Kearnes, and B. Wynne. 2005. Nanotechnology, governance, and public deliberation: What role for the social sciences? *Science Communication* 27: 268–291.

Macoubrie, J. 2006. Nanotechnology: Public concerns, reasoning and trust in government. *Public Understanding of Science* 15: 221–241.

Malsch I. (ed.) 2005. Benefits, risks, ethical, legal and social aspects of nanotechnology (2nd ed). Nanoforum General Report 4. Available at www.nanowerk.com/nanotechnology/reports/reportpdf/report3.pdf

Mao, S.S., and X. Chen. 2007. Selected nanotechnologies for renewable energy applications. *International Journal of Energy Research* 31: 619–636.

Martinez, A.W., S.T. Phillips, E. Carillho, and G.M. Whitesides. 2010. Microfluidic paper-based analytical devices. *Analytical Chemistry* 82: 3–10.

Maynard A.D. 2006. Nanotechnology: A research strategy for addressing risks. *Project on Emerging Nanotechnologies*. Available at www.nanotechproject.org/file_download/77

Maynard, A.D., D.B. Warheit, and M.A. Philbert. 2011. The new toxicology of sophisticated materials: Nanotoxicology and beyond. *Toxicological Sciences* 120(Suppl 1): S109–S129.

Mehta, M.D. 2004. From biotechnology to nanotechnology: What can we learn from earlier technologies? *Bulletin of Science, Technology & Society* 24: 34–39.

Meridian Institute. 2005. Nanotechnology and the poor: Opportunities and risks. Available at http://www.merid.org/Content/Projects/Global_Dialogue_on_Nanotechnology_and_the_Poor.aspx?view=docs

Mnyusiwalla, A., A.S. Daar, and P.A. Singer. 2003. "Mind the gap": Science and ethics in nanotechnology. *Nanotechnology* 14: R9–R13.

Monahan, T., and T. Wall. 2007. Somatic surveillance: Corporeal control through information networks. *Surveillance & Society* 4: 154–173.

Oberdörster, G., E. Oberdörster, and J. Oberdörster. 2005. Nanotoxicology: An emerging discipline evolving from studies of ultrafine particles. *Environmental Health Perspectives* 113: 823–839.

Parr, D. 2005. Will nanotechnology make the world a better place? *Trends in Biotechnology* 23: 395–398.

Pearce, J. 2012. Make nanotechnology research open-source. *Nature* 491: 519–521.

Qu, X., P.J. Alvarez, and Q. Li. 2013. Applications of nanotechnology in water and wastewater treatment. *Water Research* 47: 3931–3946.

Rai, M., and A. Ingle. 2012. Role of nanotechnology in agriculture with special reference to management of insect pests. *Applied Microbiology and Biotechnology* 94: 287–293.

Reynolds, G.H. 2003. Nanotechnology and regulatory policy: Three futures. *Harvard Journal of Law & Technology* 17: 180–209.

Roco, M.C. 2003. Broader societal issues of nanotechnology. *Journal of Nanoparticle Research* 5: 181–189.

Roco, M.C., and W.S. Bainbridge. 2001. *Societal implications of nanoscience and nanotechnology*, Report of the United States National Science and Technology Council Subcommittee on Nanoscale Science, Engineering, and Technology. Virginia: National Science Foundation.

Roco, M.C., and W.S. Bainbridge (eds.). 2003. *Converging technologies for improving human performance: Nanotechnology, biotechnology, information technology and cognitive science*, The National Science Foundation and The National Research Council. Dordrecht: Kluwer Academic Publishers.

Sabety T. 2004. Nanotechnology innovation and the patent thicket: Which IP policies promote growth? Presentation delivered at the foresight Institute 1st conference on advanced nanotechnology: Research, applications, and policy. Available at http://www.foresight.org/Conferences/AdvNano2004/Abstracts/Sabety/

Salamanca-Buentello, F., D.L. Persad, E.B. Court, D.K. Martin, A.S. Daar, and P.A. Singer. 2005. Nanotechnology and the developing world. *PLoS Medicine* 2: 300–303.

Scheufele, D.A., and B.V. Lewenstein. 2005. The public and nanotechnology: How citizens make sense of emerging technologies. *Journal of Nanoparticle Research* 7: 659–667.

Schulte, P.A., and F. Salamanca-Buentello. 2006. Ethics, nanotechnology, and the workplace: old problems, new approaches. *Environmental Health Perspectives* 115: 5–12.

Schummer, J. 2007. Identifying ethical issues of nanotechnologies. In *Nanotechnologies, ethics, and politics*, ed. H. Ten Have, 155–180. Paris: UNESCO.

Scrinis, G., and K. Lyons. 2007. The emerging nano-corporate paradigm and the transformation of agri-food systems. *International Journal of Sociology of Agriculture and Food* 15: 22–44.

Séguin, B., P.A. Singer, and A.S. Daar. 2006. Scientific diasporas. *Science* 312: 1602–1603.

Serrano, E., G. Rus, and J. García-Martínez. 2009. Nanotechnology for sustainable energy. *Renewable & Sustainable Energy Reviews* 13: 2373–2384.

Sheremeta, L., and A.S. Daar. 2004. The case for publicly funded research on ethical, environmental, economic, legal and social issues raised by nanoscience and nanotechnology (NE3LS). *Health Law Review* 12: 4–7.

Singer P.A., Daar A.S., Salamanca-Buentello F. and Court E.B. 2006. Nano-diplomacy. *Georgetown Journal of International Affairs* Winter/Spring: 129–37.

Singer, P.A., F. Salamanca-Buentello, and A.S. Daar. 2005. Harnessing nanotechnology to improve global equity. *Issues in Science and Technology* 21: 57–64.

Singer, P.A., A.D. Taylor, A.S. Daar, et al. 2007. Grand challenges in global health: The ethical, social and cultural program. *PLoS Medicine* 4, e265.

Sosnik, A., and M. Amiji. 2010. Nanotechnology solutions for infectious diseases in developing nations. *Advanced Drug Delivery Reviews* 62: 375–377.

Susanne, C., M. Casado, and M.J. Buxo. 2005. What challenges offers nanotechnology to bioethics? *Law and the Human Genome Review* 22: 27–45.

Sustainable Development Solutions Network. 2013. An action agenda for sustainable development: A report for the United Nations Secretary General. Available at http://unsdsn.org/2013/06/06/action-agenda-sustainable-development-report/

Ten Have, H. (ed.). 2007a. *Nanotechnologies, ethics, and politics*. Paris: UNESCO.

Ten Have, H. 2007b. UNESCO, ethics and emerging technologies. In *Nanotechnologies, ethics, and politics*, ed. H. Ten Have, 13–35. Paris: UNESCO.

Ten Have, H., and T.W. Ang. 2007. UNESCO's global ethics observatory. *Journal of Medical Ethics* 33: 15–16.

The Royal Society and The Royal Academy of Engineering. 2004. Nanoscience and nanotechnologies: Opportunities and uncertainties. Available at http://www.nanotec.org.uk/finalReport.htm

Thorsteinsdóttir, H., U. Quach, A.S. Daar, and P.A. Singer. 2004. Conclusions: Promoting biotechnology innovation in developing countries. *Nature Biotechnology* 22: 48–52.

Tindana, P.O., J.A. Singh, and C.S. Tracy. 2007. Grand challenges in global health: community engagement in research in developing countries. *PLoS Medicine* 4: e273.

Toumey, C. 2012. Does nanotech have a gender? *Nature Nanotechnology* 7: 412.

United Nations Educational, Scientific, and Cultural Organization (2005). Towards knowledge societies. UNESCO world report 2005. Available at http://www.unesco.org/new/en/communication-and-information/resources/publications-and-communication-materials/publications/full-list/towards-knowledge-societies-unesco-world-report/

United Nations Educational, Scientific, and Cultural Organization. 2006. The ethics and politics of nanotechnology. Available at http://www.unesco.org/new/en/cairo/social-and-human-sciences/ethics-of-science-and-technology/ethics-of-nanotechnologies/

United Nations Educational, Scientific, and Cultural Organization. 2014. Report of the international bioethics committee on the principle of non-discrimination and non-stigmatization. Available at http://unesdoc.unesco.org/images/0022/002211/221196E.pdf

United Nations High-Level Panel of Eminent Persons. 2013. A new global partnership: eradicate poverty and transform economies through sustainable development. The report of the high-level panel of eminent persons on the post-2015 development agenda. Available at http://www.post2015hlp.org/the-report/

United Nations Millennium Project Task Force on Science, Technology and Innovation. 2005. Innovation: Applying knowledge in development. Available at http://www.unmillenniumproject.org/reports/tf_science.htm

Varmus, H., R. Klausner, E. Zerhouni, et al. 2003. Grand challenges in global health. *Science* 302: 398–399.

Williams-Jones, B. 2004. A spoonful of trust helps nanotech go down. *Health Law Review* 12: 10–13.

Woodrow Wilson International Center for Scholars. 2007. Nanofrontiers. Developing story: nanotechnology and low-income nations. Available at http://www.nanotechproject.org/publications/archive/developing_story_nanotechnology/

Chapter 11
The National Bioethics Committees and the Universal Declaration on Bioethics and Human Rights: Their Potential and Optimal Functioning

Jean F. Martin

Abstract The Article 19 of the *Universal Declaration on Bioethics and Human Rights* calls for the establishment of ethics committees at various levels. There are different types of ethics committees and it is important to distinguish their missions as well as their compositions and operations: National Bioethics Committees, Clinical Ethics Committees, Research Ethics Committees, and Ethics Committees of professional healthcare associations.

This chapter deals with National Bioethics Committees (NBCs). After recalling their emergence over the last decades and the reasons for that movement encouraged today worldwide by the UNESCO Ethics Program – it presents the ways in which they promote the UDBHR and its principles, and try to implement it in practice. Based on the experience of such committees (in Switzerland in particular), the chapter elaborates on the conditions and rules necessary for NBCs to comply with the requirements of independence, multidisciplinarity and pluralism posed by Article 19 of the Declaration.

11.1 Introduction

The *Universal Declaration on Bioethics and Human Rights* (UDBHR) adopted by the UNESCO General Conference in October 2005, calls for the establishment of ethics committees at various levels, especially at the national level. Article 19 states that, "Independent, multidisciplinary and pluralist ethics committees should be established, promoted and supported at the appropriate level in order to:

J.F. Martin (✉)
Canton of Vaud Public Health Service, Lausanne, Switzerland

Swiss Public Health Association, La Ruelle 6, CH-1026 Echandens, Switzerland
e-mail: jeanmartin280@gmail.com

(a) assess the relevant ethical, legal, scientific and social issues related to research projects involving human beings;
(b) provide advice on ethical problems in clinical settings;
(c) assess scientific and technological developments, formulate recommendations and contribute to the preparation of guidelines on issues within the scope of this Declaration;
(d) foster debate, education and public awareness of, and engagement in, bioethics".

Since the adoption of the Declaration, assistance to countries in establishing National Bioethics Committees (NBCs) and training their members has been a major part of the work of the UNESCO. In the following statements issued on the occasion of 20 years of activities of its International Bioethics Committee (IBC), UNESCO elaborates its position in this regard, "… The convergence of technologies is opening new ethical, social and legal challenges for both developing and developed countries. Responsible scientific and technological innovation makes possible sustainability, ethics and social desirability. … The establishment and sharing of global bioethics norms, rules and practices is essential. …Scientific research and the advances it brings can be a key driver of development".

In order to deal with the new biotechnologies it emphasizes on the importance of addressing related ethical issues in a comprehensive way. As it says, "…however, if the ethical perspective is not taken into account at the same time as the epistemological and methodological considerations, it may not only lead to abuses of human rights, but also to inequality in economic and social development between and within countries. The development of national infra-structures in bioethics, by stimulating the establishment of national bioethics committees and promoting awareness, public debate and bioethics education for all stakeholders is a way to promote systematic decision-making" (UNESCO 2013). It should be noted that this process, from bottom to top and from top to bottom, can contribute to a greater respect for human rights by facilitating the participation of citizens in decisions that affect them. In the UNESCO view, reflection on bioethical issues is as important for scientific development as it is for economic, social and democratic developments.

Though they are often designated in short under the same name of "ethics committees", it is important to distinguish several types of ethics bodies, working in different contexts and with different missions (UNESCO 2005). Currently, there are four types of ethics committees, (1) National Bioethics Committees are entrusted with studying fundamental issues and scientific and technological developments, counselling the authorities and informing the public at large. These Committees' reports and opinions are made public. (2) Clinical Ethics Committees are responsible to examine ethical issues in a hospital or health care organizations. They provide consultation and ethical guidance to healthcare providers, organization as well as patients and their families. (3) Research Ethics Committees are affiliated to a medico-scientific infrastructure or other entity, have as their goal to make sure that scientific research protocols guarantee adequately the rights and interests of involved individuals. (4) Ethics Committees affiliated to health care professional associations. The focus of this chapter is on National Bioethics Committees.

11.2 The Importance of Bioethics Committees at the National Level

In some countries such as Switzerland, the popular "saloon bar" political feeling might be that down-to-earth common sense is enough to provide answers to the bioethical challenges and dilemmas arising from developments in biomedicine. This is similar to the ideas of those who guillotined Lavoisier during the French Revolution on the ground that "the Republic does not need scholars". This simplistic approach evidently needs to be refined.

The need for such national bodies flows from increasingly rapid scientific progress and the new possibilities opening up, which create tensions between what can and what may be done, both in terms of where and when limits should be put and in regard to the principle of justice and equitable access for all. Health care professionals, researchers, patients and society itself increasingly face crucial issues. In many cases, the traditional principles of medical ethics are not providing the necessary answers. The core mission of NBCs is to examine these issues, in an interdisciplinary effort to ascertain what constitutes responsible action at the interface of the biological sciences, medicine and health care considered in their social context at the national level.

The legislators' room for manoeuvre is often unclear, particularly in view of the complexity entailed in assessing the developments and interests involved. Law-making is a lengthy and often cumbersome process and laws are rather rigid instruments, which do not render justice to the characteristics of health care and the human relations dimension within it. Basic legal framework has of course to be developed to establish fundaments, for example in transferring in the national setting the principles anchored in the UDBHR. Further, it is a role of a high-level body like the NBC to give advice as to what should be put in the law and what might be the object of other types of rules and regulations, which might be issued by other instances. The same is true for as regards technical matters. It should be understood that Governments and Parliaments, or the Ministry of Health, shall when necessary and relevant promulgate on this basis formal public law documents (be they proper laws or regulations/ordinances/prescriptions of a lower level). The basic function of NBCs is to serve as platform for providing guidance and advice to policy makers and governments in their States. Thus, they can reinforce the role of UNESCO as an international clearing house for ethical issues and increase the audience of UNESCO work and documents. They are among the most important intermediary bodies for the implementation of the UNESCO normative instruments adopted by the member states.

In the light of the new knowledge and possibilities, the National Bioethics Committees endeavours to clarify issues and produce ethical judgements that are both clear and conducive to discussion. Doing this, it is required to represent the various conceptions on ethics and values. Its opinions are meant to foster debate and ultimately contribute to the well-being of the people concerned and of society. It doesn't provide ready-made answers, its goal is not to lay down the supposedly only politically or morally correct positions for the country, but it does make a substantial contribution to the discussion among the public and the authorities. Its advices/reports may also be aimed specifically at professionals, educational institutions or indeed economic actors, as the case may be. Thus, preparing opinions and recommendations and communicating them to the intended audiences are now at the forefront of its mission. Again, the work of the committee is not supposed to replace the legislative and political process and its recommendations are not legally binding and do not impinge on the legislative and executive powers, though possibly coming up with draft regulations. However, examining controversial situations and their ethical implications, studying possible alternative approaches forms the bulk of the committee's work and might sometimes be even more important than its final recommendations, which may not be accepted unanimously by its members or the population at large. Experience shows that the value of the contribution made by the NBC is closely linked to its diversity of viewpoints and the balance between these elements. Without limiting the necessary debate, one of its objectives should be to seek consensus positions. The more its recommendations reconcile an array of relevant ethical views on a potentially controversial subject, the sounder they are as a basis for decision-making.

The fact is that such a body, charged with advising the authorities on major and potentially explosive issues, has today an indispensable role as a source of independent advice giving to policy-makers. It works to the standards of, and in close contact with, the international debate. It is important to note that NBCs work and advice should be an aid to perceiving and evaluating the issues; it is by no means a matter, obviously, of allowing the public authorities to evade their responsibilities by transferring them to a group of State-approved "moral experts".

11.3 National Bioethics Committees: Thirty Years of Development

The need to reflect on the ethical dimension of advances in science and technology, as well as the desire to promote informed and transparent public policies that can enhance public health, has led to the establishment of NBCs, across the world. One of the first in its kind is the "US National Commission for the Protection of Human Subjects of Biomedical and Behavioural Research" (1974) that was established as part of the 1974 National Research Act. This Commission is best known for its Belmont Report, which identified fundamental principles for research involving human subjects and was the basis of subsequent federal regulation in this area (The Belmont Report 1979). In the United States, currently, the "Presidential Commission for the Study of Bioethical Issues" nominated by President Obama, continues the nearly 40-year history of groups established by the president or Congress to provide expert advice on bioethics topics. However, regarding the NBCs in the present meaning, concerned with ethical issues from a general perspective, the "French National Consultative Ethics Committee for Health and Life Sciences" (CCNE), established by President François Mitterrand in 1983, can be mentioned as one of the first bioethics committees with such mandates. This CCNE has celebrated its 30th anniversary in 2013. Its mission was defined as to provide advice on ethical problems and societal questions raised by the progress of knowledge in the fields of biology, medicine and health (CCNE 2013). The issues of medically assisted procreation and experiments on humans were the first to be addressed by the CCNE but its scope of investigation soon extended to other topics. Publishing advisory opinions is one of the CCNE's key missions. Most often, these opinions are responses to questions referred by stakeholders, for instance from the President of the Republic, Parliament, or scientific associations. However, the Committee may work on "self-refer" questions raised by the members of the Committee. The CCNE aims at encouraging members of the public to reflect on ethical matters and gain a better understanding of the topic under evaluation. The CCNE organizes an annual meeting dedicated to public debate on different bioethical issues.

In Denmark, ethical problems arising in areas such as genetic engineering, assisted reproduction and fetal examination caused the Danish Minister for the Interior to set up a committee in 1984. At that time an intense media debate about reproductive technologies captivated the public in Denmark. The committee suggested that a central ethical council should be created by law. Accordingly the "Danish Council of Ethics" was established in 1987; and began its work in 1988. In addition to questions of technologies related to human health, the "Danish Council of Ethics" gives advice and information on issues concerning nature, the environment and foodstuffs. However, the Minister of Health has no instructional powers towards the NBC and the Minister has no obligation to follow the recommendations made (Danish Council of Ethics 2014).

In the United Kingdom there is an unusual situation, where the institution giving advice at the national level emanates from a charitable trust, Nuffield Foundation. The Nuffield Council on Bioethics, an independent body that examines and reports on issues in biology and medicine, was established by the Trustees of the Nuffield Foundation in 1991. The Council's terms of reference require it mainly to identify and define ethical questions raised by recent advances in biological and medical research in order to respond to, and to anticipate, public concern; and to make arrangements for examining such questions with a view to promoting public understanding (The Nuffield Council on Bioethics 2014). As in several other countries, in the United Kingdom there are additional institutions involved in bioethics, with diverse mandates, such as the "Human Genetics Commission", the "Human Fertilisation and Embryology Authority" (HFEA 1991), the "Central Office for Research Ethics Committees" (COREC), the "National Research Ethics Service". The HFEA was established in 1991, in reflect to the birth of Louise Brown, the first "test-tube baby", in 1978. The 1984 report of the *Committee of Inquiry into Human Fertilisation and Embryology*, chaired by Lady Warnock, had recommended to establish a regulator body for human embryo research and assisted reproduction treatments.

In Switzerland, the "National Advisory Commission on Biomedical Ethics" was established by the Federal Council (Government) in 2001, under Article 28 of the

Federal Act on Medically Assisted Procreation. When the law was debated (in 1998), the Federal Parliament considered it important to have a standing ethics committee, an independent, extra-parliamentary deliberative body that would be responsible for monitoring human health issues. It should be noted that, although, the Commission is established by the Act on assisted procreation, the scope of the Swiss Commission's mandate is not limited to this field. Its mandate is to consider the whole gamut of themes and issues relevant to biomedicine, health care and human health. As NBCs in general, it does not decide, but gives advice to the authorities and the public at large. The ordinance (Governmental implementing regulation) for the Commission requires it to monitor the development of biomedical methods and techniques in research and practice and to take positions by formulating opinions on the relevant social, scientific and legal issues. In particular, the Commission is required to: keep the public informed; encourage public dialogue on ethical issues; draw up recommendations or directives for medical practice; draw attention to legislative gaps and implementation problems and, where necessary, submit proposed amendments; and advise Parliament, the Federal Council (Government) and the cantons (States) on request.

Administratively attached to the Federal Department of the Interior (which includes Offices in charge of health, health insurance, social security, culture, among other things), the Commission is competent to work independently. It makes its position papers/reports public through the media, via internet, etc., without having to obtain any particular approval from the authorities. The Swiss NBC maintains relations with equivalent committees in a number of countries, particularly its neighbours, Austria, Germany, France and Italy. It is represented at periodic meetings of national ethics committees from around the world and Europe.

11.4 NBCs and International Organizations

For more than a decade National Committees have been given much attention in the UNESCO bioethics programs. As of 2013, 17 countries had established such a body with the support of UNESCO, within the framework of the Assisting Bioethics Committees (ABC) program. So far, ten countries (most of them in Africa, plus Jamaica and El Salvador), have benefited from courses for members of their respective NBCs (UNESCO 2013).

The establishment and training of committees is facilitated by three publications (Guidelines) developed by UNESCO, which are of great help to better understand and implement the roles of NBCs; these include, *Establishing Bioethics Committees*, *Bioethics Committees at Work* and *Educating Bioethics Committees* (UNESCO 2005, 2006 and 2007). It is worth mentioning that there are other related documents such as the *Bioethics Core Curriculum* developed by UNESCO, which currently has been used in twenty universities around the world, and the first two volumes of the UNESCO Bioethics *Casebook Series* (on Human dignity and human rights, and on Benefit and harm). These documents along with the publications by UNESCO

Bioethics Chairs in several countries have made significant contributions. They are highly relevant, cover a large scope of issues in a practical and stimulating fashion, and constitute a solid basis for the teaching of physicians and other health and social professionals as well as the members of NBCs.

With the help of UNESCO Programs, more and more often, there are contacts between NBCs from a given region. Those committees in the European Union meet quite regularly to discuss topics of common interest. Other multinational meetings bring together NBCs linked by a common language. Over the last 20 years, about half of the nations of the world have followed suit and created their own NBCs, which however are at various levels of institutionalization and activity. As of 2013, the World Health Organization (WHO) lists about a hundred countries with such committees. As the Permanent Secretariat, WHO convenes a Global Summit of National Bioethics Advisory Bodies every two years, to discuss the relevant issues (WHO Global Summit 2014).

11.5 NBCs and the UNESCO Declaration on Bioethics and Human Rights

The National Bioethics Committees are an important instrument to contribute to the greater awareness of the *UNESCO Declaration on Bioethics and Human Rights* (UDBHR). The NBCs are a canal and leverage for the dissemination of the principles embodied in the Declaration. For instance, regarding the aims of the Declaration, Article 2 states, "to foster multidisciplinary and pluralistic dialogue about bioethical issues between all stakeholders and within society as a whole"; and Article 18, about Decision-making and addressing bioethical issues, recommends that; (1) Professionalism, honesty, integrity and transparency in decision-making should be promoted, in particular declarations of all conflicts of interest and appropriate sharing of knowledge. Every endeavour should be made to use the best available scientific knowledge and methodology in addressing and periodically reviewing bioethical issues. (2) Persons, professionals concerned and society as a whole should be engaged in dialogue on a regular basis. (3) Opportunities for informed pluralistic public debate, seeking the expression of all relevant opinions, should be promoted.

Also in terms of Article 23, which emphasizes bioethics education, training and information, and recommends that, "in order to promote the principles set out in this Declaration and to achieve a better understanding of the ethical implications of scientific and technological developments, in particular for young people, States should endeavour to foster bioethics education and training at all levels as well as to encourage information and knowledge dissemination programmes about bioethics", NBCs have a critical role in the implementation process of this recommendation.

A review of Article 19 of the Declaration explains the structure and function of such committees. It directs that, in order to fulfil their mandate adequately, the committees shall be independent, multidisciplinary and pluralist. Based on the experience of the Swiss National Committee (Martin 2009, 2012), as well as those in other countries, a number of "good practice" elements can be identified with a view to ensure the successful work of such bioethics committees at the a national level.

Given the role of NBCs, it is essential to form a properly balanced group of people with recognized ethical standards who, while holding and capably defending strong opinions of their own, can practice dialogue and cooperation. Such committee is not a parliament, where controversial or doctrinaire figures might be very visible and disruptive, but a group of "sages" (wise persons) working in concert on complex issues. Thus, committee members' capacity for dialogue is as important as the discipline in which they have their background. The UNESCO Guide No. 2 for national bioethics committees states this clearly: "Once the role of the Committee has been determined, it must be filled in order to function. The quality of membership will obviously be crucial in determining its success. Well-chosen members can often make even badly designed institutions work; poorly chosen members can doom even the best designed structure" (UNESCO 2006, p. 22).

11.6 NBC's Structure, Composition and Function

The official document instituting the committee will have provisions on its composition and the disciplines or characteristics it should embody. These include experts from medicine, biomedical research, public health, nursing, law, ethics, philosophy, theology, sociology and psychology, and possibly economics. A few members might be drawn from the general population as laypersons (civil society, patient representatives etc.). The number of committee members varies. It depends on the function of the committee; for example, if the work is to be carried out in plenary sessions, then there are usually between 15 and 20 members. However, if much preparatory activity is done by working groups and plenaries are rarer, the NBC might be larger, for example, the French CCNE has 39 members while, smaller than most, the US Presidential Commission for the Study of Bioethical Issues presently has only 11 members. While, in a committee with a large membership, there is a risk that interminable debates will render them ineffective, in a small committee the membership and the views expressed may not be pluralist enough. The desirable structure for the committee membership is that the members be appointed in a personal capacity, even if they owe this appointment to their membership of a particular profession or spiritual group, for example. They should know that they are not "delegated" by the groups to which they belong and are not required to defend -in a "unionist" fashion- the interests of these groups. Their mission is, after listening to the views of others, to make their personal contribution to the discussions and recommendations of the committee in the general interest, without being accountable to anyone else. The expression of professional, political or religious opinions is wholly admissible, but not in the form of dogmatic pro domo pleading.

This being so, having ex officio appointments is questionable because persons nominated in this way will be bound to see themselves as delegates of a certain group or authority and will lose some freedom as well as public accountability. Nor is it desirable for members to be co-opted; this procedure is unlikely to ensure the necessary diversity of opinion or genuine independence of the members. The matter of the term of appointments needs to be settled. It is important to appoint members for a limited periods, e.g. of 4 years, perhaps renewable once or twice. It does not seem advisable to have life appointments (even if in theory the assurance that they will remain in their post at all events is a guarantee of independence). There is a risk that an institution which is very slowly renewed and is ineluctably ageing would become ossified and lacking in initiative and originality. For the NBCs it is important to be clear about issues such as incompatibilities between NBC membership and other duties. In Switzerland for example, neither the members of the National Parliament nor federal civil servants may be members of extra-parliamentary national commissions such as the Bioethics Advisory Commission. In brief, the membership is to consist of individuals from various backgrounds and talents. In addition to their specialized knowledge, a prime qualification is that they be widely seen as thoughtful people who are good listeners, have considerable experience of life and are capable of operating on an interdisciplinary basis, debating constructively with colleagues in different fields. There must be a gender balance as well as a balance between philosophical and religious views and the main sections of society in the country.

The NBC must be genuinely independent, both intellectually and practically (including budget wise), so that it operates without undue restraint or oversight. Similarly, it does not require prior authorization – imprimatur – of other authorities before making its positions public. The issue of "conflict of interest" is an increasingly present and sometimes acute cause for concern, rightly so, in relation with the composition and work of public authorities and other institutions, and in the debates on civil and political processes. It is essential that there be as much transparency and disclosure as required about any ties that could trammel committee members' independence and limit their freedom to express their true opinion. There must be a certainty that they will not behave as lobbyists for any interest group. In terms of working topics, the official document establishing the NBC may require it to produce studies on particular questions. The committee may work on topics and questions requested by the specific authorities such as parliament or ministers. Further, it must be free to take up issues that it considers important. It is also desirable that it is accessible for members of the public and ordinary citizens. In these cases, the NBC may, but is not obliged to, take up the matter if it considers that the enquiry is relevant and falls within its remit.

Bioethics is a very broad field. Even if it is desirable for few or no restrictions to be set on the topics to study, priorities must be established, as the resources, time availability of members, and funding, are limited. Logically, one would expect the NBC to concentrate on bioethical issues that are of particular importance to its country, in its present circumstances or in the future. It is worth monitoring that if there is a report or document developed by other national bioethics committees, elsewhere, on a similar question, which might be relevant and useful for the country, and thus can be adopted, possibly subject to adjustments. Such endorsement of opinions issued in other countries contributes to the emergence of positions that are shared as widely as possible around the world. Careful consideration should be given to the practical consequences of the opinions required from the NBC. Therefore, situations in which people may experience unnecessary suffering or refusal of specific benefits until the opinion of the NBC is known should be dealt with due diligence.

In dealing with bioethical issues and dilemmas, sometimes there are problems for which there is no clear cut answer or an optimum solution. The committee should then have the courage to go for the best option available (which might be the "less bad"). In case of diversity of opinion, controversy and disagreement among the committee members notwithstanding, it is highly recommended to reach a consensus. This may be achieved, by formulating a position that everyone can support without being forced to betray his/her own convictions and can often be done through thorough, measured discussion. If consensus cannot be reached, the commission will decide whether it is advisable to publish an opinion despite diversity of views. If the committee hopes to win the ear of the authorities and public, it must formulate recommendations that are as clear and practical as possible. These recommendations are advisory only, and the authorities will be less willing to follow them if the NBC's statements are vague or lacking in coherence and direction, or if they have failed to convince a clear majority of the committee. The committee's opinions and recommendations should in principle be made public and widely available for all.

Based on UNESCO recommendation, self-education is crucial for the members of Ethics Committees. UNESCO in its Guide no 3. *Educating Bioethics Committees*, underlines that "Experience, in short, has refuted the old assumptions that life had sufficiently prepared members for their task or that their pre-existing moral and social values rendered them impervious to change, or that self-education by committees was at best redundant" (UNESCO 2007, p. 9). In its Guide No. 1. *Establishing Bioethics Committees*, it also emphasizes that, "Members of NBCs may be persons of distinction, but few are experts in all the areas of their committee's purview – and fewer still are learned in bioethical inquiry. One of the members' main tasks, then, becomes self-education. Much of this proceeds informally – members learn from each other, talk with knowledgeable outsiders and canvass existing literature. Some self-educations, however, are formal" -, such as holding seminars, and inviting external experts for lectures (UNESCO 2005, p. 22). However, there is a real challenge here. The difficulty for committee members to keep up, even in their own discipline, with the very rapid developments of the biomedical field, in its various dimensions, leading to unexpected situations, must not be underestimated. Thus it is definitely not always easy to exchange and work in an inter-disciplinary and trans-disciplinary manner. In all fields, including bioethics, there is a need for ongoing training to create familiarity with new ideas and practices and to effectively follow the new developments. NBC members must have these opportunities, including easy access to the international literature and contacts. At the simplest, parts of its sessions will be given over to presentations by experts of matters of interest.

The committee's effectiveness depends heavily on its secretariat. It is the mainspring of its operations, working closely with the chairperson or executive committee, and under his or her responsibility. It should be re-emphasized that it is important for them to be independent practically and intellectually, even when the NBC is attached to a ministry for administrative purposes. The secretariat ensures continuity of the committee's activities. It must have capabilities in the bioethical field broadly defined, be able to routinely work in an interdisciplinary fashion and to ensure that contributions from different sources are used productively. It is most useful for it to have, in addition to management and administrative competence, skills in the drafting of technical papers and other texts. Needless to mention, in order to carry out its functions, the NBC and the Secretariat must have available resources, human as well as technical, adequate to their tasks.

11.7 Conclusion

What place should peoples be giving to ethical thinking and practice in the early twenty-first century? Ethical commitment is much needed in the biomedical sciences and health care, and the UNESCO Universal Declaration on Bioethics and Human Rights provides a solid framework in this regard. It is equally essential to keep in mind the larger global-wide problems such as climate change, global water supply, food security as well as efforts to achieve progress towards a sustainable and equitable world. Beyond their specific mandates, National Bioethics Committees have a significant role to play in mobilizing the attention and commitment of governments and societies in order to confront the dangers threatening our common future. This raises the difficult but necessary question of "global governance". That socio-economic circumstances and opportunities remain so dissimilar in various parts of the world is one of the major ethical challenges. Though solutions are eminently difficult to find, it should be remembered that, from any point of view, these differences are unacceptable and, as often underlined in discussions within the International Bioethics Committee, this is a concern for global bioethics. Every country, however modest in size and importance, must make it a duty to contribute effectively to the national and international debates. It is a matter of urgency to focus on the ethical basis of our policies and actions, which caused the current global environmental and financial crises. Paradigm shifts seem essential. Based on their characteristics of independence, multidisciplinarity and pluralism, national bioethics committees can prepare the ground for such changes, in their field of expertise.

References

CCNE. 2013. Available at: http://www.ccne-ethique.fr/en. Last visited 26 Feb 2014.
Danish Council of Ethics. 2014. Available at: http://etiskraad.dk/en.aspx. Last visited 26 Feb 2014.
Human Fertilisation and Embryology Authority (HFEA). 1991. Available at: http://www.hfea.gov.uk/25.html. Last visited 26 Feb 2014.

Martin, J. 2009. Le travail d'un comité national de bioéthique – Questions de principe et de pratique. *Bulletin des Médecins Suisses (Journal of the Swiss Medical Association)* 90: 438–441.

Martin, J. 2012. Des principes importants pour l'éthique clinique. *Bulletin des Médecins Suisses.* 93: 302.

The Belmont Report. 1979. Available at: http://www.hhs.gov/ohrp/humansubjects/guidance/belmont.html. Last visited 26 Feb 2014.

The Nuffield Council on Bioethics. 2014. Available at: http://nuffieldbioethics.org/. Last visited 26 Feb 2014.

UNESCO. 2005. Guide No. 1. Establishing bioethics committees. Paris: UNESCO. Available at: http://unesdoc.unesco.org/images/0013/001393/139309e.pdf. Last visited 26 Feb 2014.

UNESCO. 2006. Guide no 2. Bioethics committees at work: Procedures and policies. Paris: UNESCO. Available at: http://unesdoc.unesco.org/images/0014/001473/147392e.pdf. Last visited 26 Feb 2014.

UNESCO. 2007. Guide no 3. Educating bioethics committees. Paris: UNESCO. Available at: http://unesdoc.unesco.org/images/0015/001509/150970e.pdf. Last visited 26 Feb 2014.

UNESCO. 2013. 1993–2013: 20 years of bioethics at UNESCO. Paris: UNESCO. Available at: http://unesdoc.unesco.org/images/0022/002208/220865e.pdf. Last visited 26 Feb 2014.

WHO Global Summit. 2014. Available at: http://www.who.int/ethics/globalsummit/en/. Last visited 26 Feb 2014.

Chapter 12
The UNESCO International Bioethics Committee and the Network of Ethical Advisory Bodies in Europe: An Interactive Relationship

Christiane Druml

Abstract Since its creation in 1993 the International Bioethics Committee (IBC) has developed important declarations in the field of bioethics shaping the worldwide debate with a focus on goals that are important for all UNESCO Member States. The IBC members are independent experts representing different regions, professions as well as different cultures of the currently 195 UNESCO member states. The UNESCO declarations do not constitute legal sources by themselves and consequently are not legally binding, but exercise an important influence on bioethical debates. However, the more specific the topic and the "newer" the topic in the biomedical research area, the more probable will be its adoption in other international or national documents.

As the bioethical debate in the United States and Europe is historically strong, the influence of the IBC and its documents should be examined in a different way compared to other regions. However, with respect to Europe, one has to distinguish between the various geographical definitions of Europe: the European Union, the Council of Europe and UNESCO. An important influence of the IBC in Europe comes from the activities of European members of the committee in academia and the various national ethical bodies.

12.1 Introduction

The bioethics debate worldwide has been driven by issues of medical research involving human beings. One of the cornerstones in this debate has been the Nuremberg Code (1947) and its requirement for "informed consent" of persons participating in a medical research project (Shuster 1997). Since then, the autonomy

---

C. Druml (✉)
Medical University of Vienna, Vienna, Austria
e-mail: christiane.druml@meduniwien.ac.at

of human participants in biomedical research projects has been respected and officially acknowledged. In parallel we have witnessed the development of structured ethical review of medical research. In 1964 the World Medical Association established guidelines governing medical research in the Declaration of Helsinki(2013). Later in 1978 the first amendment of the Declaration introduced "Ethics Committees" -as independent bodies- to review medical research protocols. Thus the ethical review system which today is an integral part of clinical research all over the world was created. Today (Research) Ethics Committees have to be established in institutions where medical research on human beings is conducted to ensure ethical review of clinical research projects. While the name of the committee might differ, for instance, Institutional Review Board, Ethical Review Committee or Independent Research Ethics Committee, its mandate is generally the same: to review research protocols to see whether or not the integrity and wellbeing of patients or healthy volunteer participants are protected. These committees are composed of experts, scientists and lay persons. Physicians, nurses, lawyers, pharmacists, patient representatives and ethicists provide necessary expertise on these committees.

In Europe there are many different laws and "soft laws" governing biomedical research. In the EU, requirements for the conduct of clinical trials are provided by the "Clinical Trials Directive" (2001/20/EG of the European Parliament and of the Council of 4 April 2001 on the approximation of the laws, regulations and administrative provisions of the Member States relating to the implementation of good clinical practice in the conduct of clinical trials on medicinal products for human use). The EU Directives have to be transformed into national law within the single European Member States (Druml 2009). Further revision of European law for clinical trials has been adopted by regulation which is directly applicable in all EU countries (Regulation No 536/2014 of the European Parliament and of the Council on clinical trials on medicinal products for human use, and repealing Directive 2001/20/EC). The regulation entered into force on 16 June 2014 but will apply no earlier than 28 May 2016. This means that the Clinical Trials Directive will still apply until that date.

It should be noted that the function of Research Ethics Committees in Europe is based on the same legal requirements and the success of these committees has paved the way for the establishment of advisory Bioethics Committees in Europe (Huriet 2009). While Research Ethics Committees have the mandate to review specific, clinical research projects on human beings, Bioethics Committees are established at the local, regional or national level. Their mandate is to advise a body such as, a government or a house of parliament, on general ethical questions arising from the advancement of biomedical research. When their mandate is at the national level, they are called "National Bioethics Committees". Furthermore, in general they also have other obligations like fostering public debate and awareness in those issues, and making recommendations for specific laws and regulations. The work of Bioethics Committees is assisted by other similar bodies established within professional organizations, with a focus on advising their organizations on bioethical matters of importance and specific cases within their mandates (UNESCO 2005).

In Europe the French "Conseil Consultatif National d'Ethique" (CCNE) was the first National Bioethics Committee founded (CCNE 1983). In February 1983, after the research summit "Assises de la recherche", France's President François Mitterrand issued a decree creating the first National Consultative Ethics Committee for Health and Life Sciences. Conducting human experiments was one of the first topics to be addressed by the CCNE, along with ethical issues surrounding medically assisted procreation. Many other European States have followed France's lead by establishing national advisory bodies for bioethical issues. They differ in composition and working method, but are more or less independent. The members are selected because they are considered the experts in influential fields such as law, medicine, natural sciences, philosophy, ethics and political sciences. However, in Europe there are also other non-national advisory bodies, which provide advice and recommendations for regulations on bioethical topics.

12.1.1 European Group on Ethics in Science and Technology

In Europe there is a body specifically established to advise the European Commission. This committee, which has been named "The European Group on Ethics in Science and Technology" (EGE), is also independent, pluralistic and multidisciplinary and was originally established in 2001 by the European Commission (EGE 2001). Its mandate was renewed in December 2009 and the 15 members from European Member States advise on the ethical aspects of science and new technologies with respect to preparation and implementation of legislation and policies. The committee works either on its own initiative or following a request of the European Commission. Parliament and Council are in the position to draw the Commission's attention to questions which they consider to be of major ethical importance. The Bureau of European Policy Advisers, (BEPA) is the office of the European Commission responsible for bioethics and ethics of science and new technologies and serves as the secretariat of the European Group on Ethics in Sciences and New Technologies (EGE). BEPA is tasked with disseminating the recommendations and decisions of EGE group. What is furthermore important is that BEPA represents the European Commission in meetings on bioethics and ethics of science organized by relevant third parties like UN agencies, the Council of Europe, International Organizations, and nationally and internationally relevant authorities in this field. BEPA also organizes the European Commission's international dialogue on bioethics, where the EGE, the Chairs of 15 non-EU National Ethics Councils from 5 continents, the Chairs of the EU National Ethics Councils as well as representatives of international organizations meet to share information, discuss the main bioethics topics and create synergies. This is one of the occasions where a representative of the International Bioethics Committee of UNESCO presents its current work to a large expert audience within the European area.

12.1.2 National Ethics Councils Forum

Networking has high priority on the agenda in any European ethical discourse. The European Science in Society Program (SIS) hosts the Forum of National Ethics Councils (NEC Forum). This Forum, established in 2001 by the Council of Ministers, is independent and serves as an informal platform for the exchange of information and best practices on issues of common interest in the field of ethics and science (NEC 2001). There are two meetings every year, hosted by the country that holds the current EU Presidency. Participants in these meetings are the chairpersons and secretaries of National Ethics Councils. These meetings also include a joint session with the European Group on Ethics thus ensuring exchange of information and high-level ethical advice to the European Commission. To widen the information and exchange, representatives from the Council of Europe, UNESCO and WHO are routinely invited to attend the joint meetings.

12.1.3 Council of Europe

Another player in the field of bioethics is the Council of Europe. Already in the year 1985, very early in the history of bioethics in Europe, the Committee of Ministers had set up an Ad hoc Committee of experts on bioethics. This Ad hoc Committee was responsible for intergovernmental activities of the Council of Europe in the area of bioethics and was later transferred to the Steering Committee on Bioethics (CDBI). It was this Committee which developed the "Convention for the Protection on Human Rights and Dignity of the Human Being with Regard to the Application of Biology and Medicine: Convention on Human Rights and Biomedicine". The Convention was the first international treaty in the area of Bioethics, adopted by the Committee of Ministers in the year 1996. The Convention, also known as the "Oviedo Convention" from the city where it was signed, came into force in 1999. It sets out the fundamental principles applicable in day-to-day medicine as well as those applicable to new technologies in human biology and medicine. Currently the Convention has been signed by 34 of the 47 Member States of the Council of Europe and 29 Member States have ratified the Convention (Oviedo Convention 1999). Among the European Member states which have not yet signed are Austria, Germany, Belgium and the United Kingdom.

The Convention addresses the minimum standard of biomedical activities on which there is a consensus among European countries. A main topic in the Convention is research on persons unable to give consent. Another main issue is research on the embryo. However, it should be noted that for some countries the minimum standard is too permissive, while for others it is the opposite (Druml 2009). It should be noted that there are several additional Protocols, the Protocol on the Prohibition of Cloning Human Beings, the Protocol concerning Transplantation of Organs and Tissues of Human Origin, the Protocol on Biomedical Research, and the Protocol concerning Genetic Testing for Health Purposes. Today the Committee on Bioethics (DH-BIO) has taken over the responsibilities of the Steering Committee on Bioethics (CDBI). This was done after a reorganization of the intergovernmental bodies at the Council of Europe in January 2012 (DH-BIO 2012).

12.2 UNESCO Bioethics and Europe

UNESCO is the only United Nations agency with a specialized mandate in the social and human sciences, which means that UNESCO is a frontrunner in the field of bioethics and is getting more important due to the exponential growth of biomedical research and its consequences to humankind. The documents issued by the International Bioethics Committee (IBC) of UNESCO are the results of global consensus and provide an ethical framework for policies on sensitive bioethical issues. An example is the field of genetics and the human genome. This area is particularly noteworthy as UNESCO was able to focus on this topic early while the issue was still "new", and its impact was influential as researchers began building biobanks to store bodily fluids and tissues (Salter and Jones 2005).

However the influence of the IBC's documents is limited to a certain extent as its documents are issued in the form of 'soft' law. 'Soft' law is not legally binding, but is obviously preferentially chosen in the field of bioethics. This is also due to the fact that the role of states in the field of bioethics is more about promotion than implementation (Andorno 2007; Boussard 2009). In fact, the IBC with its 36 independent members from various disciplines and all regions illustrate UNESCO's global membership. It should be noted that the Intergovernmental Bioethics Committee (IGBC) has also been established by the UNESCO member states with the mandate to link the decisions of the IBC with the activities of the specific national governments. The IGBC was created in 1998 under article 11 of the Statutes of the IBC and its 36 members represent their government's views and opinions on bioethical issues. The task of members is to examine the advice and recommendations developed by the IBC based on the opinion of their respective governments. The European representation in both committees, IBC as well as IGBC, provides a platform to present European perspectives on different bioethical questions under consideration by UNESCO. It should be noted that the notion of "Europe" is different within the various organizations. The European Union includes of today 28 European Member States. The Council of Europe entails a much wider group of national countries: 47 countries and 6 observer states. The notion of the European region within UNESCO encompasses again a wider array of nations, namely 53 members and one associate member.

Since its creation, and as required by its Statutes, the IBC has a balanced geographical representation and therefore experts from the European region -some of them being chairs of National Ethics Committees or holding important positions in academia- have always played an important role in the IBC. In the first period from 1993 to 1995, more than half of its members represented Europe. Currently, in

2013, 11 of the 36 experts are from Europe. Furthermore in the year 2008 a majority of 67 % of the members were also members of their National Bioethics Committee and 14 % of those were the chair. This reflects a particularly intense interweaving with national committees (Vöneky 2010) thus helping to shape the bioethical debate within the single states. Since 1993, UNESCO's role is to actively and continuously shape the bioethical debate in a global context through its universal declarations and documents and even more so through its various ethics programs such as the Assisting Bioethics Committees (ABC) program and other initiatives.

The UNESCO programs have been very instrumental in bioethics capacity building in some parts of Europe. For instance, the ABC program has influenced Europe particularly in regions like the Russian Federation, and Ethics Teacher training Courses (ETTC) have been held in mainly Eastern European countries such as Romania, Slovakia, Lithuania, Serbia, Azerbaijan and Croatia. In the case of the EU Member States, due to the fact that Western Europe has traditionally been strong in developing its own law and soft law in this field, the influence of the IBC is less visible compared to non-Western European regions. Europe also is the place of origin of other important national bioethics organizations such as the Nuffield Council in the UK, Deutscher Ethikrat in Germany, and *CCNE* in France, as well as other regional bodies such as the European Group on Ethics in Science and Technology.

The UNESCO chairs in Bioethics, which have been created to assist the implementation of the bioethics programs, are very active in some parts of Europe such as Italy, Portugal (established in 2009), Spain (established in 2007), and in Presov, Slovakia (established in 2010).

The main influence of the International Bioethics Committee however, can be seen in the activities of the European members themselves and consequently in the involvement of the European IBC members in their own national UNESCO Commissions. It should be noted that the National Commissions for UNESCO in each country are very important agencies who implement IBC documents in their countries. On the other hand, as mentioned earlier, the IBC members are also members of national bioethics committees which advise politicians and government bodies early on about the new developments in the ethics of science and technology. Therefore, bioethics activities in the five UNESCO program areas- Education, Natural Sciences, Social and Human Sciences, Culture, Communication and Information- receive more national visibility. It is important to note that other United Nations agencies are also active in bioethical discussions and policy making. For instance the World Health Organization (WHO), which is a global authority for health is active in the field of health ethics and has developed various guidelines, basic principles and standards. The Department of Ethics, Equity, Trade and Human Rights as well as the Department of Knowledge, Ethics and Research, coordinate bioethics-related activities in the World Health Organization.

The WHO also hosts the permanent secretariat for the Global Summit of National Ethics Committees (WHO 1996), and manages the database of worldwide existing National Ethics Committees and the database called Opinions of National Ethics Committees (ONEC) – which is publicly accessible. Supporting this endeavor, the

Global Network of WHO Collaborating Centers for Bioethics has been established to help implement WHO ethics mandate.

In 2003, another bioethics body, the UN Inter-Agency Committee on Bioethics (UNIACB) was established at the initiative of UNESCO. This committee has been serving as a key inter-agency mechanism for sharing information between and among intergovernmental organizations dealing with issues related to bioethics, thus fostering better cooperation and coordination at the international level. Several UN Agencies such as Food and Agriculture Organization (FAO), International Labor Organization (ILO), Office of the U.N. High Commissioner on Human Rights (UNHCHR), UN Educational, Scientific, and Cultural Organization (UNESCO), World Intellectual Property Organization (WIPO) and World Health Organization (WHO), are members of this committee and UNESCO serves as the permanent secretariat of the committee.

References

Andorno, R. 2007. The invaluable role of soft law in the development of universal norms in bioethics. Deutsche UNESCO Kommission, Juli 2007. Available at: http://www.unesco.de/1507.html. Last visited 25 Jun 2014.
Boussard, H. 2009. Role of states. In *The UNESCO Universal Declaration on Bioethics and Human Rights; Background, principles and application*, ed. H. Ten Have and M.S. Jean. UNESCO Publishing.
Committee on Bioethics. DH-BIO 2012. Available at: http://www.coe.int/t/dg3/healthbioethic/cdbi/default_en.asp
Conseil Consultatif National d'Ethique. 1983. Available at: http://www.ccne-ethique.fr/en. Last visited 20 May 2014.
Council of Europe. Ad Hoc Committee on bioethics. Available at: www.coe.int. Last visited 20 May 2014.
European Commission Responsible for Bioethics and Ethics of Science and New Technologies. Available at: http://ec.europa.eu/bepa/european-group-ethics/bepa-ethics/index_en.htm. Last visited 20 May 2014.
European Group on Ethics in Science and New Technologies. 2001. Available at: http://ec.europa.eu/bepa/european-group-ethics/welcome/index_en.htm. Last visited 20 May 2014.
EU Directives. 2011. Available at: http://ec.europa.eu/health/human-use/clinical-trials/index_en.htm. Last visited 20 May 2014.
Declaration of Helsinki. 2013. Available at: http://www.wma.net/en/30publications/10policies/b3/index.html. Last visited 20 May 2014.
Druml, C. 2009a. Research ethics committees in Europe: Trials and tribulations. *Intensive Care Medicine* 35(9): 1636–1640.
Druml, C. 2009b. Stem cell research: Toward greater unity in Europe? *Cell* 139(4): 649–651.
Huriet, C. 2009. Article 9, Ethics Committees. In *The UNESCO Universal Declaration on Bioethics and Human Rights; Background, principles and application*, ed. H. Ten Have and M.S. Jean. UNESCO Publishing
National Ethics Councils Forum. Available at: http://ec.europa.eu/research/science-society/index.cfm?fuseaction=public.topic&id=1305. Last visited 20 May 2014.
Oviedo Convention. 1999. Convention on Human Rights and Biomedicine. Available at: http://conventions.coe.int/Treaty/Commun/ChercheSig.asp?NT=164&CM=&DF=&CL=ENG. Last visited 20 May 2014.

Salter, B., and M. Jones. 2005. Biobanks and bioethics: The politics of legitimation. *Journal of European Public Policy* 12(4): 710–732.

Shuster, E. 1997. Fifty years later: The significance of the Nuremberg Code. *The New England Journal of Medicine* 337: 1436–1440.

UNESCO. 2005. Guide No. 1, UNESCO, Establishing Bioethics Committees. Available at: http://unesdoc.unesco.org/images/0013/001393/139309e.pdf. Last visited 20 May 2014.

United Nations Inter-Agency Committee on Bioethics. 2003. Available at: http://en.unesco.org/events/13th-meeting-united-nations-inter-agency-committee-bioethics-uniacb#sthash.felrUGMC.dpuf. Last visited 20 May 2014.

Vöneky, S.R. 2010. Moral und Ethik. Grundlagen und Grenzen demokratischer Legitimation für Ethikgremien; 198.

WHO. 1996. Available at: http://www.who.int/ethics/globalsummit/en/. Last visited 20 May 2014.

Chapter 13
The Impact of the UNESCO International Bioethics Committee's Activities on Central and Eastern Europe

Olga Kubar and Jože Trontelj

Abstract The impact of the UNESCO International Bioethics Committee's (IBC) initiatives over its 20 years in the region of Central and Eastern Europe (CEE) is an excellent example of a mutually beneficial two-way process: the great influence of the IBC on bioethics capacity building at national and regional levels, as well as the contribution of 15 bioethics experts from CEE countries to the IBC's activities as members of the committee. Due to the great historical and economical changes that took place in the regional countries over the past 20 years, there have been some unique impacts by the IBC on the CEE countries. During the process of implementing universal bioethical principles proclaimed in UNESCO IBC Declarations, CEE countries have focused on communicating the necessity and challenges of achieving independence, competence, openness and responsibility in the field of bioethics. Development of regional and international cooperation facilitates free discussions, the exchange of experiences, as well as capacity building in bioethics.

13.1 The Impact of UNESCO-IBC Declarations and Initiatives on the CEE Countries

The process of forming the Commonwealth of Independent States (CIS) gave rise to a unique experience of dynamic legislative, administrative and informational development in the sphere of ethical regulation in biomedicine. This new concept of multilateral cooperation united 11 regional countries (The Azerbaijan Republic, Republic of Armenia, Republic of Belarus, Georgia, Republic of Kazakhstan,

---

Sadly Prof. Jože Trontelj passed away before the publication of this book.

O. Kubar (✉)
Clinical Department, Pasteur Institute, Saint-Petersburg, Russia
e-mail: okubar@list.ru

J. Trontelj
Slovenian Academy of Sciences and Arts, Ljubljana, Slovenia

Kyrgyz Republic, Republic of Moldova, Russian Federation, Republic of Tajikistan, Republic of Uzbekistan and Ukraine). As it was recognized that consensus in bioethics discussions could not be reached without first understanding and respecting political, economic, socio-cultural, historical, and religious differences, a detailed review of these aspects was carried out in a study supported by the UNESCO Moscow Office and the UNESCO Headquarters in the CIS countries. The study showed that "bioethics initiatives and consolidating activities in law-making, education, creation of the system for the ethical review and international cooperation have become a priority" (Kubar et al. 2007). The information obtained in this study was important for the dissemination, promotion, elaboration and application of ethical principles of the UNESCO Declarations as well as for bioethics programs, such as: Assisting Bioethics Committees, Ethics Education Programs and Global Ethics Observatory in the CEE countries (UNESCO Bioethics Programs 2013). It should be noted that in the successful realization of these goals, the role of UNESCO regional offices was very important. Following the creation of the UNESCO Assisting Bioethics Committees program, National Committees on Bioethics were established in all cluster countries of the UNESCO Moscow office, including: Azerbaijan, Armenia, Belorussia, Moldova and Russian Federation and in many other CEE countries. Special attention was focused on the networking and capacity building of these Committees to foster the exchange of information, support decision-making, develop tools for standard setting, and strengthen coordination among experts and institutions in the region (Volik 2010; Petrov and Yudin 2010). The concept of an Ethics Education Programme was inspired by the general historical goal of achieving ethical practice in the sphere of healthcare, based on universal ethical values. In fact the achievement of this goal depended upon the establishment of international cooperation for the implementation of the Ethics Education Program as a measure for promoting UNESCO Universal Declarations. As a first step for implementing the Ethics Education Program in the region, based on the experience of bioethics education in Azerbaijan, Armenia, Belarus, Georgia, Kazakhstan, Kyrgyz Republic, Moldova, Russian Federation, Tajikistan, Uzbekistan and Ukraine, an analytical review of the current state of bioethics education in these countries was conducted (Kubar et al. 2010; Kubar 2013). A great contribution has also been made by the UNESCO Ethics Teacher Training Courses, in which students from more than 10 countries in the region and regional experts on bioethics took part. The positive results of such activity include the contributions of the Polish Bioethics Society and the Ukraine Bioethics Association at the national level, and the sub-regional Association for Education in Bioethics in the Commonwealth of Independent States (CIS) at the international level. Following Ethics Teacher Training Courses, educational bioethics programs at the regional level were organised to increase capacities in the area of ethics education and to improve all activities in the fields of bioethics and human rights. It should be noted that the UNESCO's Ethics Education Program was translated into many national languages and pilot testing was conducted in local Universities. Accordingly, bioethics became a mandatory subject for students in biological and medical schools and there are courses at the Master's level in countries such as Georgia, Poland and Russia. Activities with a special focus to promote ethical principles and to raise public awareness about bioethical issues as well as introduce bioethics into the agenda of the mass media for journalists were carried out by the UNESCO Moscow office in the long term project entitled "Bioethics and Media" (Tishenko and Yudin 2008). The importance of bioethics education has been emphasized by the establishment of the Units of the International Network of the UNESCO Chair in Bioethics in 10 regional countries: Albania, Armenia, Azerbaijan, Bulgaria, Croatia, Czech Republic, Macedonia, Russia, Serbia and Ukraine. This activity was supported by regional political bodies to facilitate the global process of bioethics capacity building and to facilitate bioethics-related legislation. The participation of the CEE region in the Global Ethics Observatory, in all its six databases provides a platform for experts' collaborations and exchange of information at the regional and global levels. The need for bioethical guidelines is growing world-wide and has been addressed by several international bodies. In Europe one should mention the contributions of at least two organizations in this regard: the Bioethics Committee of the Council of Europe (DH-BIO) and the European Group on Ethics in Science and New Technologies (EGE). However, as the work of these bodies partly overlapped with that of UNESCO, some small countries like Slovenia mainly took part in the regional rather than global activities. In terms of the IBC's impact on bioethics development in this region, special attention should be paid to the influence of UNESCO Declarations in the capacity building of human resources in bioethics. Another important development is the impact of the Universal Declaration on the Human Genome and Human Rights (UNESCO 1997) and the *International Declaration of Human Genetic Data* (UNESCO 2003) on regional legislation by the Recommendation of Inter-Parliamentary Assembly of the CIS "Ethical and Legal Regulation in Genetic Research in the CIS States", and on national legislation in many regional countries (Kubar 2010; Petrov and Yudin 2010). The impact of the Universal Declaration on Bioethics and Human Rights (UNESCO 2005), which introduces ethical principles on important topics such as the protection of vulnerable populations, non-discrimination and non-stigmatisation on the development of bioethics discourse in this region, is noticeable. It should be noted that the UNESCO Declarations have been translated into all regional languages and are available on the UNESCO related websites and other media resources. These documents have become part of the professional and public educational programs in bioethics.

13.2 Bioethics in the CEE Region: Future Development and Challenges

More than ever, humankind needs to consider the place of ethics in its future. UNESCO and its International Bioethics Committee are well positioned to address ethical, legal and social issues in biomedicine and to stress the importance of equitable distribution of benefits, protection of human rights, and the fight against discrimination and poverty in sustainable development. One cannot be satisfied with the present world-wide level of respect for ethics. Selfish interests still underlie most human activities, as manifested in the struggle for power and wealth. We are still witnessing horrible abuses with massive human rights violations, tragic wars, genocides, poor governance of natural and human resources, and a sadly limited capacity of the United Nations for efficient interventions.

Even in countries enjoying peace, relative welfare and democracy, we can see decreasing respect for human dignity, justice and equity. This is somehow the result of giving priority to material interests over human values. This may be a dangerous track to follow. We see potentials for new misuses of the achievements of science and technology. Current philosophical ethical debate shows a cross-roads questioning the validity of the concept of human dignity. However, the necessity to revisit and re-define human rights is evident. It is highly appropriate to have this kind of debate at UNESCO and other international agencies such as the World Health Organization (WHO) or the World Medical Association (WMA). However, in this regard the regional bodies such as the Council of Europe and European Group for Ethics in Science and New Technologies as well as the various national bodies can play a great role. These activities provide several types of top-down approaches, such as recommendations for the national legal systems, social policies, etc. However, the way towards improving function at the global level without changing human behavior is unimaginable, not only at the collective but also at the individual level. This will critically depend on a greater acceptance of ethical values in personal decision making, i.e. in individual conduct. In other words, there is an indispensable need for a bottom-up approach. Building a better society means reinforcing moral sensitivity and increasing awareness of personal ethical responsibility through the complete range of individual roles embodied in each person. The most efficient way to improve personal acceptance of ethical values is by reinforcing school education in ethical values (Carr and Steutel 1999). This needs to be strengthened at all levels, from pre-school to the university. The aim should not be limited just to achieving elementary literacy in ethics. The main goal should be to stimulate value-based reasoning and choices. UNESCO with its global mandates seems to be well positioned to prepare appropriate recommendations and help the national bioethics committees towards that goal. These recommendations should take into account the new knowledge in neurosciences on the developing brain (Mustard 2006), developmental medicine and child psychology as well as pedagogy. A significant amount of work has already been done by UNESCO, but there is still much to be accomplished.

Commitment to coherent, progressive, properly structured education in values for all young people may produce far-reaching beneficial effects upon the life of future societies. Therefore it is our suggestion to apply the established as well as recently accumulated knowledge on teaching and learning to a framework of carefully prepared educational programing of cultural, social and political values. What our societies really need is responsible behaviour by all members at all levels of societies. A long-term effort to strengthen school education in ethical values would be of paramount importance.

During the last 20 years a dynamic process of development has taken place in the CEE countries. Capacity building in bioethics is part of these developments, and the activities of the UNESCO International Bioethics Committee had a great impact in shaping bioethics discussion in this region. The implementation of the universal principles of bioethics introduced by UNESCO creates a unique opportunity to analyze the extent of such impact on the bioethics discourse in this region. It is noteworthy that cooperation in this sphere is bilateral and multilateral. There is no doubt that cooperation between the UNESCO International Bioethics Committee and the national bioethics committees in the CEE region can be enhanced by more participation of bioethics experts from IBC member countries. Such collaboration will facilitate the implementation of UNESCO Declarations and recommendations to increase respect of human dignity, rights and freedoms, equal access to scientific achievements, flow and exchange of knowledge and mutual benefits.

References

Carr, D., and J. Steutel (eds.). 1999. *Virtue ethics and moral education*. London: Routledge.

Kubar, O. 2010. *Bioethics in the CIS countries: Engaging in ethical discourse*, 108–112. National Bioethics Committees in Action, UNESCO.

Kubar, O. 2013. *20 years of IBC UNESCO in the aspect of the regional development bioethics*, 96. LAP LAMBERT Academic Publishing is a trademark of: OmniScriptum GmbH and Co. KG Heinrich-Böcking, Germany.

Kubar, O., G. Mikirtichian, and A. Nikitina (eds.). 2007. *Ethical review of biomedical research in the CIS countries (Social and cultural aspects)*. Saint-Petersburg: United Nations Educational, Scientific and Cultural Organization.

Kubar, O., G. Mikirtichian, and A. Nikitina (eds.). 2010. *The current state of bioethics education in the system of medical education in the CIS member countries (analytical review)*, 200. Saint-Petersburg: Saint-Petersburg Pasteur Institute.

Mustard, J.F. 2006. Experience-based brain development: Scientific underpinnings of the importance of early child development in a global world. *Paediatrics & Child Health* 11(9): 571–572.

Petrov, R., and B. Yudin. 2010. *The development of bioethics in Russia*, 18–26. National Bioethics Committees in Action, UNESCO.

Tishenko, P., and B. Yudin. 2008. *Bioethics and media: Recommendations for journalists*, 58. Moscow: Adamant.

UNESCO. 1997. Universal Declaration on the Human Genome and Human Rights. Available at: http://www.unesco.org/new/en/social-and-human-sciences/themes/bioethics/human-genome-and-human-rights/. Last visited 10 Jan 2014.

UNESCO. 2003. International Declaration on Human Genetic Data. Available at: http://www.unesco.org/new/en/social-and-human-sciences/themes/bioethics/human-genetic-data/. Last visited 10 Jan 2014.

UNESCO. 2005. Universal Declaration on Bioethics and Human Rights. Avaialble at: http://www.unesco.org/new/en/social-and-human-sciences/themes/bioethics/bioethics-and-human-rights/. Last visited 10 Jan 2014.

Volik, B. 2010. *The national medical ethics committee of Slovenia*, 53–55. National Bioethics Committees in Action, UNESCO.

Chapter 14
Bioethics in Arab Region and the Impact of the UNESCO International Bioethics Committee

Sadek Beloucif and Mohamed Salah Ben Ammar

Abstract Arabic countries are in the Eastern Mediterranean region which is the cradle of the three major monotheistic religions Judaism, Christianity and Islam. While Arabic countries share a religious and cultural background, a recognizable diversity exists among these countries due to various schools of jurisprudence. Since its establishment, the UNESCO International Bioethics Committee (UNESCO-IBC) has had very strong ties with Arabic countries. This chapter examines how the UNESCO IBC has shaped and impacted bioethics development in the Arab region. It also describes how Arabic countries, in turn, have contributed to international bioethical debates.

14.1 Introduction

The Arab region has its own cultural identity and traditions. This unique situation is based on the unity of language, culture, religion, socio-economic circumstances and demographic distribution. Even though Arabs feel they belong to the same community and share common values, they are not completely homogeneous. Coarsely we can divide Arabic countries into three big entities, even if the boundaries of these regions do not represent the true divisions between the entities: Gulf countries, North-African countries and Middle Eastern countries. Gulf countries include: Bahrain, Kuwait, Oman, Qatar, Saudi Arabia, United Arab Emirates and Yemen. North-African countries include: Mauritania, Morocco, Algeria, Tunisia, Libya, Egypt and Sudan. Middle Eastern countries include: Lebanon, Syria, Palestine, Iraq and Jordan. In terms of bioethics development, it is important to mention that not all these countries are on the same level. For instance, among these countries, Kuwait,

S. Beloucif
Anesthesiology and Critical Care Medicine, Sorbonne Paris Cité University, Paris, France

M.S. Ben Ammar (✉)
Anesthesiology and Critical Care Medicine, El Manar University, El Manar, Tunisia
e-mail: msbenammar@gmail.com

Qatar, Mauritania, Morocco, Palestine and Iraq have no ethics committees (Abou-Zeid et al. 2009). The contribution of Muslim physicians to medical sciences and practice is well documented and reflected in the medical literature especially between ninth and sixteenth centuries. A review of the flourishing time of Islamic medicine in medieval times shows a great attention to the ethical issues in medicine by the Muslim physicians such as Rhazes (865–925 AD), and Avicenna (980–1038). For centuries onward, their medical ethics instructions were dominant teaching and practice in medicine which followed by other prominent Muslim physicians (Bagheri 2014). In the contemporary era, however, development of the Islamic Code of Medical Ethics (1982) by the International Organization for Medical Sciences (IOMS), in 1981 in Kuwait is noticeable (IOMS 1981). In the Arab world based on Islamic teachings, bioethics is applied in a different context than in the West. While there are Shiites, Sunnis and other minorities, in Arabic countries, bio-ethical reflections are based on Islamic *Shari'a*. Despite this common foundation however, a diversity of approaches in bioethics exists in Arabic countries. This diversity derives from the various schools of jurisprudence, the different sects within Islam, the differences in cultural background as well as the different levels of religious observance (Daar and Al Khitamy 2001).

In addition, values that would be considered universal at the global level may not be fully adopted by some religious communities, indicating that they are not always concerned by secular bioethics (Filiz 2011). Currently, many Arabic countries, have established local ethics committees at their universities and hospitals, independent ethics committees (IEC), or research ethics boards (REB), which are all very active. In addition, National Ethics Committees (NEC), which was recommended by UNESCO, have been established in many countries in the region. Although, the NEC's missions are completely different from those of local committees, it is not uncommon to see a local committee playing the role of NEC, which may create confusion and tension. However, several problems still exist at the national level: national ethics committees do not meet regularly; and their composition is not as diverse as that recommended by UNESCO especially in terms of gender balance. In Arab region, the contribution of scientists and physicians in bioethical discussions is dominant, and the participation of anthropologists, lawyers, philosophers and sociologists is not as sought after as it should be. Furthermore, national ethics committees only have an advisory role with no real authority (Abou-Zeid et al. 2009). An added complexity is that many prefer to talk about Muslims, not about Islam; considering that not all Muslims are Arabs, and not all Arabs are Muslims. In this chapter we use these two terms interchangeably as we believe that this better reflects the reality of our society.

14.2 Ethics and Islamic Teachings in the Arab Region

For many Muslims, religious beliefs are a fundamental part of both personal and social existence in their daily life and a major determinant for healthcare decision-making. In Islamic teachings, the importance of inter-human as well as human-divine relations has been emphasized. The prophet of Islam, Muhammad (puh), announces the perfection of morals as the aim of his appointment. He says, "I have been appointed as the prophet for the completion and perfection of morals" (Abu Abdallah Muhammad 1002 AD). In the field of bioethics, UNESCO aims to provide a platform where ethics, science, culture, philosophy and medicine are joined together towards a common good. The general aim is to associate, and not oppose, all cultures to create a virtuous circle for an "osmosis of cultures", recognizing that all individuals are equal in dignity. At the global level, there is a growing awareness of the ethics work of Muslim scholars, bioethicists, physicians and philosophers. For the Arabic mind, the concept of healthcare implies involvement not only of healthcare providers, but also of legal and/or religious experts within a framework of social responsibility. Based on Islamic teaching, the concept of health is associated with hygiene, hydration, comfort and care. Its boundary, based on the global concept of physical, psychological, social and spiritual well-being extends to interacting with the patients and their family in the context of community. This approach forms a cornerstone of the duties of the physician which necessarily includes attention to respect of the person and ethics in general. It is stated in the Holy Qur'an: "We have honoured the children of Adam" (17:70). Other potentially relevant citations would be: "God enjoins equity and charity" (16:90); "My Lord enjoins fairness" (7:29); "Be fair; God loves those who are fair" (49:9).

Ethical considerations in Arab region are not restricted to the medical field and encompass good manners and morals. Furthermore discussions cannot be confined to a battle between autonomy and collective community interests. In light of this tension, human dignity is a complex concept that has been proposed to be at the core of bioethics. This has been emphasized in several international documents such as the *UNESCO Universal Declaration on the Human Genome and Human Rights* (1997). However, it can be found as early as in the opening of the Preamble of the 1948 *Universal Declaration of Human Rights*, adopted by the United Nations, where it states that: "Whereas recognition of the inherent dignity and of the equal and inalienable rights of all members of the human family is the foundation of freedom, justice and peace in the world" as well as in its first Article where it says: "All human beings are born free and equal in dignity and rights. They are endowed with reason and conscience and should act towards one another in a spirit of brotherhood" (United Nations 1948). Dignity however is difficult to clearly define. Like the principles of bioethics, it is rooted on practical grounds to prevent exploitation and abuse, but is also aiming for clear respect for principles and values. In fact, one of the difficulties (and beauty) of dignity is that it possesses a dual acceptance. Taken as a general principle, it infers that it would be a means to protect against external aggressions, against the excessive "liberty" of others on the one hand, and on the other hand, it would be understood as an individual claim, a means to promote an individual conception of liberty. However, a question remains whether dignity is more related to an individual, or to a group value. The question is very relevant to Arab philosophy. Medieval and modern Arab philosophers have taught us that if we consider justice as an equivalent of dignity (with the goal of achieving equity and equality of chances, and then fighting against discrimination), we find a harmonious agreement between different theories of justice: Procedural (libertarianism);

Utilitarianism; or Egalitarianism. In 2004 in a conference in Paris, the International Bioethics Committee brought together representatives of various world religions. The conference concluded that it is possible to formulate universal principles based on common values. It also acknowledged the existing differences in moral views on particular issues. In this regard reference was made to the Muslim religion exemplifying a common ethic among so many different cultures, nations, and traditions (ten Have and Gordijn 2013). When Western philosophy says: "I think, therefore I am"; the East responds: "You are, therefore I am". Such an understanding can be useful in the light of a trans-cultural approach. Given the individuality and singularity of humans, we are all different. However, to enhance global understanding, a constructive cross-cultural dialogue should be promoted.

International ethical guidelines, such as the Nuremberg Code, Helsinki Declaration, WHO Guidelines, and reports of UNESCO in the field of bioethics, have been developed and are applied in many countries including Arabic countries. However, cultural characteristics still need to be taken into account more carefully. These cultural differences explain, for example, why the issue of regulating stem cell research elicited concerns from some scientists reluctant to see any regulation on stem cell, while other regulating bodies tried to balance scientific freedom with ethical constraints. Another example is the issue of prohibitions of human reproductive cloning, and the production of hybrid forms combining human germ cells and the germ cells of other species. In the globalized world, these kinds of research do not only simply "belong" to the West, and this type of experimentation concerns all mankind, medically, scientifically, culturally, socially, anthropologically, and even spiritually. Faced with the new advances of science, each citizen can consider ethical deliberations as a method of protection which demands justification before continuing with the scientific development. As the predominant ethical value shifts from "beneficence" to "autonomy", the application of biomedical sciences has evolved progressively from a purely paternalistic approach to a contract-based approach. We should in fact act as humanists, promoting a true therapeutic alliance between the scientists and the public. This evolution will provoke (and already does) specific tensions to the Arabic countries where traditionally "the good of the group" often prevails over "the good of the individual" as the value of autonomy becomes less relevant than in the West.

In Islamic teachings, individuals are responsible for their health, and have a duty to preserve it. However, physicians are also professionally responsible for ill individuals in their societies, which is clearly stated in medical oath. It worth mentioning that the Hippocratic Oath is acknowledged in Islamic bioethics however, invocation of multiple gods in the original version of the Hippocratic Oath, and the exclusion of all gods in later versions, has led Muslims to adopt the Oath of the Muslim Doctor, which invokes the name of Allah (Daar and Al-Khitamy 2001). For Muslims, a physician is believed to have a strong and special statute. He is the *Hakim*, a person full of wisdom and knowledge. In practice, he is able to respond to patients' suffering, expectations, and to build confidence and respect. However, patients should not remain passive, with the development and progress of modern medicine, a total (or blind) trust in physicians is not recommended. It has been argued that moral judgement belongs to everyone (Thomas 1999), therefore, participation of patients in healthcare decision making is very important. In Islamic societies, in everyday clinical practice, physicians have to make choices for the good of their patients and he should pay attention not only to the "how" of his medical practice, but also the "why". That is, he has to act as a true professional, diligently respecting ethical values and duties. Medical situations are always multi-faceted and cannot yield to automatic answers given the complexity of the many ethical elements that have to be considered.

In Islamic society, careful attention to ethics is expressed through various initiatives such as; the annual sessions of the Council of the Islamic Academy of Fiqh; fatwas issued by authorized Islamic scholars as well as the Council of Islamic Medical and Scientific Ethics (IMSE) (2009–2012); Islamic Organization of Medical Sciences (IOMS); the Arab League Educational Cultural and Scientific Organization (ALECSO); the Islamic Code of Medical and Health Ethics (EM/RC52/7 September 2005).

The need for dedicated ethics training is recognized and several organizations promote strategic initiatives to enhance bioethics capacity in this region (Eckstein 2004). This, together with the creation of ethics committees, generates a reasonable optimism that bioethical topics identified as priorities will be addressed in the coming years. Several studies have focused in priority settings in Islamic bioethics. In 2007, at a UNESCO meeting in Cairo the following topics were ranked as very important in Islamic countries: Cloning; Stem cell research; Genomic ethics; Study of the human genome; Research on human organ transplantation; Reproductive technologies; Pharmaceutical research; Medical practice and abortion.

In terms of research ethics, the following topics were reported as being very important for the national bioethics committees in this region: monitoring of biomedical research; assessment of understanding of informed consent; privacy and confidentiality; provision of appropriate risk reduction measures; assessment of cultural sensitivity for informed consent; placebo controlled trials; determination of appropriate subject selection in vulnerable populations; assessment of anticipated benefits; community participation; determinations to conduct phase I, II and III clinical trials in a country or community; and Incentives to participation (Abou-Zeid et al. 2009).

In another effort, an international questionnaire has listed the top 10 bioethical challenges in the Muslim world as: The relationship between law, ethics and *fatwa;* Justice and health resource allocation; Human rights; Bioethics capacity-building; Patients' rights; Brain death and organ transplantation; Individual autonomy and informed consent; Islamic principles of bioethics; Abortion; and Bioethics committees (Bagheri 2013).

14.3 Bioethics Development in the Arab Region

In the Arab region, the Islamic Organization of Medical Sciences (IOMS) has been one of the first institutions that analyzed Medical Ethics and Bioethics from an Islamic perspective. Beginning from its establishment in 1981 this organization has organized several international conferences on Islamic Medicine that have embraced a wide range of topics with special emphasis on elaborating the Islamic perspectives regarding the recent technological developments in biomedicine and medical practice. In 1981, this organization developed a key document entitled the *Islamic Code of Medical Ethics* in its first international conference in Kuwait. Since then the organization has sponsored several regional meetings and has published significant works in this field. To address the challenges posed by globalization, the seventh Conference in Cairo (2002), was dedicated to the topic of "the impact of globalization on development and health services in Islamic countries" (IOMS 1981).

In March 2002, the international conference on healthcare ethics was held in Abu-Dhabi, United Arab Emirates, to promote cross-cultural dialogue on bioethics and to explore the possibility of universal standards in healthcare ethics. The conference explored the ways to increase the contribution of the Islamic countries in global bioethics (Conference Proceedings 2002).

The same year, the Organization of the Arab League for Education, Science and Culture (ALESCO) established an advisory committee entitled "Arab Committee for Ethics of Science and Technology". The first meeting of this committee took place in Beirut, Lebanon in August 2003, during which its general strategy, objectives and work plan were discussed and approved. Parallel to the meeting, scientific symposiums on "Ethical aspects of assisted human reproduction" and "medical ethics as a teaching discipline in medical schools" were organized (Hattab 2003). Another important regional development which deserves to be highlighted for its international impact was the creation of the Network of Islamic Bioethics by the International Association of Bioethics almost a decade ago. Currently, in Arabic countries, there are ethics committees, research centers and institutions which function at the national level. For instance, Saudi Arabia hosts the Committee of Ethics on Medical Practice in the Gulf States and in this country there is also an Ethics Committee on Brain Death and Persistent Vegetative States. In Egypt, the International Islamic Center for Population Studies in Al-Azhar University has been a pioneer center working on ethical issues in medicine, especially reproductive health. The first bioethics committee in Egypt was formed in Al-Azhar University in 1991, and this committee developed a bioethics curriculum in 2000. In the United Arab Emirates, the Gulf Center for Excellence in Ethics was established in 1998. In Kuwait, the Center for Organ Transplantation of Ibn Sina (Avicenna) Hospital provides ethics guidance in organ transplantation. In Lebanon, Salim El-Hoss Bioethics and Professionalism Program in American University of Beirut is an active bioethics center. The Saint Joseph University also has a graduate diploma in Ethics, open for specialists in natural and social sciences and humanities.

In Yemen, bioethics initiatives in the University of Aden provide undergraduate as well as post graduate bioethics programs (Hattab and Jose Ramon 2010). Compare to other bioethical topics, ethics in research and research ethics committees in Arab region have been more discussed in the literature. For instance, a study on ethical codes in biomedical research in 13 Arab countries shows that most of the research ethics documents in use in this region demonstrate numerous deficiencies especially in regard to the development, structure, content, and reference to international guidelines in these documents (Alahmad et al. 2012).

Like other region, in Arab region there is a continuing discussion on whether ethics committees should act as counsellors for the prince, or instead should be a committee of citizens for its citizens. With the willingness of equating ethics and morals, the overall aim would be to express an idea of promoting good will and democracy. Ethics could, in a sophisticated society, be seen as a tool to limit the risks that can arise if excessive powers are given to a group of people or professionals. For instance, in the national ethics committees in the Eastern Mediterranean region which includes Arabic countries, the activities of NECs are often reduced to review of medical research protocols (67 %); an advisory role to policy makers (47 %); training (47 %); and publications (67 %) (Abou-Zeid et al. 2009). It should be noted that in the field of bioethics other regional organizations such as the Eastern Mediterranean Regional Office (EMRO) of the World Health Organization have also promoted ethical approaches to health science and practice in the Arab region. In 1995 during the 42nd session of the Regional Committee for the Eastern Mediterranean, the EMRO technical paper on "Ethics of Medicine and Health" was developed. This report demonstrated that the cardinal ethical principles of medicine and health, as understood at the present time, form an integral part of the principles of Islam as a way of life. These principles include respect for human dignity, justice and beneficence which are very clearly expressed and emphasized in the Holy Qur'an. The committee recognized that there is a need for a detailed code of healthcare ethics to guide the countries of the region. Further, the meeting emphasized the need for more collaboration between organizations involved in bioethical debates. In 2005, at the 52nd EMRO session in Cairo, bioethical issues such as the doctor-patient relationship, and relationships between healthcare providers and community were discussed (EMRO 2005).

In 2007, the First Regional Meeting of National Bioethics Committees was held in Cairo, at the WHO Eastern Mediterranean Regional Office (EMRO). The meeting was organized jointly by the UNESCO Cairo office and EMRO, and was the first time the two international organizations had collaborated in holding such a meeting in the region. Experts from 15 countries in the region took part, representing Member States from both the UNESCO Arab Region and the WHO Eastern Mediterranean Region. In this meeting several senior technical staff participated from the Arab League Educational, Cultural and Scientific Organization (ALECSO), the Gulf Cooperation Council (GCC), and the Islamic Organization for Medical Sciences (IOMS).

During this meeting, the Cairo resolution developed a set of recommendations including: urging countries which do not yet have an established ethics committee to establish a national bioethics committee with help and support from UNESCO and WHO; setting up a network for communication between existing committees; compiling an Arabic bioethics dictionary; sharing bioethics education for medical and science students; as well as raising public awareness of ethics so that people know their rights. It is important to note that in this region, like other societies, the interaction between individual rights and community interest is an important and relevant issue.

14.4 Arab Region and UNESCO Bioethics Initiatives: Of Mutual Benefit

Like in other regions, the interactions between a geographic area and an international institution such as the UNESCO-IBC are complex. The beauty of specific interactions between UNESCO's IBC and various regions at the global level is that they will tend to unite humans while respecting cultural diversity. These fruitful interactions are useful tools to improve transcultural dialogue by referring to common humanity. It has been argued that the rising awareness of the global dimension of bioethics has major impacts on Islamic bioethics. These impacts can be distinguished at two different levels: the global and local. At the global level, bioethics is advancing a transcultural framework of ethical values and principles in which Muslim scholars have contributed much to the international effort to identify global values and principles that are commonly shared among all human beings (Brockopp and Thomas 2008).

A review of different bioethics initiatives in Arab region indicates that how recommendations made by international organizations, such as UNESCO as well as the World Health Organization, have helped Arab countries to develop and enhance their bioethics programs locally. In the last two decades, UNESCO-IBC has initiated many activities in the field of bioethics. It is noteworthy that the UNESCO Intergovernmental Bioethics Committee (IGBC) has also provided a platform for engaging member states in its bioethics programs and recommendations. For instance, in the Arab region, the IGBC was very instrumental in initiating a series of efforts to implement UNESCO's follow-up action after the adoption in 2005 of the Universal Declaration on Bioethics and Human Rights. This included dissemination of the Declaration by providing information, and publication of its translation into local languages.

Since the establishment of the UNESCO International Bioethics Committee in 1993, there have been always several members from Arabic countries. This has provided the opportunity to initiate interactive dialogue on ethics topics. This mutual collaboration has benefited both the IBC and regional members. The IBC members from Arab region have tried to implement the recommendations developed by UNESCO and also have worked to promote a global understanding of cultural diversity and values. In 2005, after 2 years of intense discussion and negotiation, all member states of UNESCO reached consensus on the text of the *Universal Declaration on Bioethics and Human Rights*. The Declaration is in fact the first international legal, though non-binding, instrument that comprehensively deals with the linkage between human rights and bioethics (Andorno 2007). As observed by Robert Veatch, this document is a "convergence of various ethical systems creating a single normative framework and speaking for virtually all citizens of the world" (Veatch 2012). Muslim countries have been well represented in the international bioethical discussions in UNESCO and its International Bioethics Committee. For instance, as members in UNESCO, during the international discussions and consultations concerning the *Universal Declaration on Bioethics and Human Rights*,

Islamic countries including Arabic countries contributed to its development. In its bioethics initiatives UNESCO has focused on ethics capacity-building to develop and enhance national bioethics committees. In practice, these committees play a great role in implementing UNESCO's various declarations and recommendations in this field (UNESCO 2014). It should be acknowledged that the following UNESCO bioethics programs were also very helpful in bioethics capacity building in this region: (1) The Ethics Education Program (EEP), which aims to reinforce and increase the capacities of member states in the area ethics education. (2) The Ethics Teacher Training Course (ETTC) also supports and promotes quality ethics education around the world by building professional capacities of bioethics teachers. In this region, this course was held in Saudi Arabia (2007), Jordan and Oman (2014) and will be held later this year in Lebanon. (3) Another program, the Assisting Bioethics Committees (ABC), which was created on the recommendation of the *Universal Declaration of Bioethics and Human Rights*, advocates the establishment of independent, multidisciplinary and pluralistic ethics committees at national or institutional levels. This program has played a very important role in providing instruction and encouraging Arabic countries to establish such ethics committees. Collaborations between UNESCO and national bioethics committees in Arabic countries have been instrumental in organizing educational programs and conferences in the region. For instance, in 2001, an important bioethics regional event "International Symposium on Ethics of Science and Technology" was organized by The National Lebanese Commission of UNESCO in cooperation with UNESCO (Lebanon National Commission 2001). Despite the successes of UNESCO's efforts in the Arab region, there is still much to be done. In 2015 the Global Ethics Observatory (GEObs), which is a system of databases with worldwide coverage in bioethics and other areas of applied ethics in science and technology, noted that in terms of bioethics experts, institutions and bioethics related legislations and guidelines, the Arab region is still behind the rest of the world. Currently, bioethics has to deal with an increasing number of topics such as ethical implications of economic constraints; health equity and resource allocation limitations, privacy and autonomy (e.g. biometry, databases, terrorism, mandatory vaccinations) as well as environmental ethics.

There is no doubt to deal with these issues, like other regions in the world, Arabic countries utilizes extensive bioethics resources developed by UNESCO-IBC to enhance bioethics discussions and policy in their countries.

14.5 Conclusions

In the field of ethics of science and technology, UNESCO is well positioned to set bioethics norms and policies, build capacity, and to increase public awareness. It also helps link experts and scientists to policy makers. In Arab countries, following UNESCO bioethics initiatives, the number of national bioethics committees has increased over the past decade. However, there are still a few countries in this region which have not yet established an ethics committee at the national level. In addition, existing committees suffer from structural deficits and lack of resources. Furthermore, a diversity of opinion about the application and interpretation of the UNESCO Bioethics Declaration exists. The establishment of bioethics institutions in Arabic countries is an opportunity to promote bioethics in the region. In this regard there are two main concerns: first is the challenge between the local (religious) and the global approach to bioethics; and the second is how to accommodate the diversity of opinions and approaches in bioethical debates in order to implement the UNESCO Declaration on Bioethics and Human Rights. Cultural and religious constraints sometimes inhibit any debate. Members of some Arabic countries still have misgivings around conducting pluralistic debates because of the variety of sensitive issues raised thereby. For these reasons NECs in Arabic countries should reaffirm that the core bioethical principles which have been emphasized by the UNESCO Bioethics Declaration are in harmony with Islamic values.

References

Abou-Zeid, M., M. Afzal, and H.J. Silverman. 2009. Capacity mapping of national ethics committees in the Eastern Mediterranean Region. *BMC Medical Ethics* 10: 8. doi:10.1186/1472-6939-10-8.
Abu Abdallah Muhammad 1002 CE (393 AH). *Al-Mustadrak alaa al-Sahihain*, vol. 2, 282.
Alahmad, G., M. Al-Jumah, and K. Dierickx. 2012. Review of national research ethics regulations and guidelines in Middle Eastern Arab countries. *BMC Medical Ethics* 13: 34. doi:10.1186/1472-6939-13-34.
Andorno, R. 2007. Global bioethics at UNESCO: In defence of the Universal Declaration on Bioethics and Human Rights. *Journal of Medical Ethics* 33: 150–154.
Bagheri, A. 2013. Priority setting in Islamic bioethics: Top 10 bioethical challenges in Islamic countries. *Asian Bioethics Review* 6(4): 391–401.
Bagheri, A. 2014. Bioethics in Iran: Development, issues and challenges. In *Biomedical ethics in Iran: An application of Islamic bioethics*, ed. Alireza Bagheri. Christchurch: Eubios Ethics Institute.
Brockopp, J.E., and E. Thomas (eds.). 2008. *Muslim medical ethics: From theory to practice*. Columbia: University of South Carolina Press.
Conference Proceedings 2002. UAE. Healthcare ethics: A cross-cultural dialogue on the ethical challenges of healthcare. *Conference Proceedings*. Abu-Dhabi, 2002: 21.
Daar, A.S., and B. Al-Khitamy. 2001. Bioethics for clinicians: Islamic bioethics. *Canadian Medical Association Journal* 164(1): 60–63.
Eckstein, S. 2004. Efforts to build capacity in research ethics: An overview. *Science and Development Network*. Available at: http://www.sci-dev.neen/middle-east-and-north-Africa/policy-briefs/
EMRO. 2005. EM/RC52/R.10. Available at: http://applications.emro.who.int/Library/Databases/wxis.exe/Library/
Filiz, S. 2011. The place of bioethics principles in Islamic ethics. In *Islam and bioethics*, ed. A. Berna and Vardit Rispler-Chaim, 23–45. Ankara: Ankara University Press.
Hattab, A.S. 2003. Report on the first meeting of Arab Committee for the Ethics of Science and Technology, Beirut.
Hattab, A.S., and A. José Ramón. 2010. Bioethics in the Arab world: The experience of Aden School of Medicine. *Revista Latinoamericana De Bioethica* pp. 8–10.

Islamic Charter of Medical and Health Ethics. 1982. EM/RC52/7 September 2005.
International Organization for Medical Sciences. 1981. Available at: http://www.islamset.com/ioms/pricelis.html
Lebanon National Commission for Education, Science and Culture. 2001. Report on ethic of science and technology. Beirut.
ten Have, H., and B. Gordijn. 2013. Global bioethics: Transnational experiences and Islamic bioethic. *Zygon* 48(3). Also see: Eleventh session of the International Bioethics Committee of UNESCO. Paris: UNESCO. http://unesdoc.unesco.org/images/0013/001395/139549e.pdf
The Islamic Code of Medical Ethics. 1982. *World Medical Journal* 29(5): 78–80.
Thomas, J.P. 1999. La bioéthique à l'épreuve de la finitude. In *La bioéthique est-elle de mauvaise foi?* vol. 8, ed. F. Diderot, 30–44. Paris: Presses Universitaires de France.
UNESCO. 2005. Universal Declaration on Bioethics and Human Rights. http://www.unesco.org/new/en/social-and- human-sciences/themes/bioethics/bioethics- and-human-rights/
UNESCO Ethics Program. Available at: http://www.unesco.org/new/en/social-and-human-sciences/themes/bioethics/ethics-education-programme/. Also see: UNESCO. 2014. Assisting Bioethics Committees. Available at: http://www.unesco.org/new/en/social-and-human-sciences/themes/bioethics/assisting- bioethics-committees/
UNESCO Global Ethics Observatory. Available at: http://www.unesco.org/new/en/social-and-human-sciences/themes/global-ethics-observatory/
Veatch, R.M. 2012. *Hippocratic, religious, and secular medical ethics: The points of conflict.* Washington, DC: Georgetown University Press.

Chapter 15
The Impact of the UNESCO International Bioethics Committee on Latin America: Respect for Cultural Diversity and Pluralism

Claude Vergès De Lopez, Delia Sánchez, Volnei Garrafa, and Andrés Peralta-Corneille

Abstract In Latin American bioethics, cultural diversity and pluralism is a critical issue. Due to the importance of human rights in Latin America, bioethicists in this continent worked hard to apply and implement the UNESCO Universal Declaration on Bioethics and Human Rights. Currently, social bioethics and research ethics are important themes of congresses, educational programs and publications. However, the main topic is bioethics and health. The aim of Latin American Bioethics is to provide the countries, Latin Americans and people in the Caribbean with an additional new instrument for improving democracy, citizenship and human rights, which is derived from the construction of an expanded and more politicized concept of Bioethics. This also includes promoting a wide trans-disciplinary exchange of information and experiences, at regional as well as international levels.

15.1 Introduction: Cultural Diversity and Pluralism in Latin America

The history of Latin America is intimately tied to the recognition of the importance of cultural diversity and pluralism after a long period of colonial domination and denial of Indigenous and African participation in the process of national

---

Sadly, Dr. Peralta passed away before the publication of this book.

C. Vergès De Lopez
Deontology and Bioethics, Faculty of Medicine, University of Panama, Panama, Panama

D. Sánchez (✉)
School of Medicine, Universidad de la República, Montevideo, Uruguay
e-mail: dibarsan@adinet.com.uy

V. Garrafa
University of Brasilia, Brasilia, Brazil

A. Peralta-Corneille
Santiago Technological University, Santiago De Los Caballeros, Dominican Republic

© Springer International Publishing Switzerland 2016
A. Bagheri et al. (eds.), *Global Bioethics: The Impact of the UNESCO International Bioethics Committee*, Advancing Global Bioethics 5,
DOI 10.1007/978-3-319-22650-7_15

empowerment. The development of bioethics scholarship in Latin America has been influenced by this historical context and coincides with the work of the bioethics unit of UNESCO and of the International Bioethics Committee (IBC) on these themes. In the specific field of health, Latin American bioethics emphasizes fairness, equity, comprehensiveness and quality of health services to comply with human rights in health. For this purpose, it has been necessary to develop educational programs in bioethics and UNESCO has been both a source of knowledge and a support.

The recognition of cultural diversity in Latin America has been the product of political, anthropological and social work with the different Indigenous populations, their struggle and the struggle of Afro-American groups for the recognition of their human rights in the middle of the twentieth century. Next, women fought for their rights from a gender perspective and currently homosexual groups are demanding non-discriminatory treatment and equity. Inequities although social class remains the main divide in the most unequal region of the World, Indigenous and/or African descendants are grossly overrepresented in the lower classes in most countries. The colonization of Latin America has caused traumas in societies, especially the denial of the human qualities of Indigenous peoples. The best intentioned approaches were justified by the values of beneficence and protection, similar to what would be applied to children and individuals with diminished capacities, and ignored the autonomy of these individuals and populations. Following the South American wars of independence, Indigenous groups or nations were relegated and forgotten. Then the works of Levy-Strauss and Latin-American anthropologists demonstrated the different social organization of Indigenous groups and their distinctive world view, bringing new interest for these groups. However, the greatest obstacle for the recognition of their human rights has been their historical loss of independence.

Latin America is considered as a whole entity yet there are large cultural differences among countries, despite the many similarities in history and dominant language. It is possible to distinguish some larger groups of cultural unity: México and Central America, the Latin Caribbean group (Spanish speaking islands, the Atlantic coast from Florida, United States to Venezuela), Brazil, the Southern Cone and the Andean countries. Although there are differences within these communities, cultural similarities have impacted Latin American societies and permitted the recognition and respect of the different groups. For Mexico, Central America and the Andean countries, the problem of equality and justice for the indigenous communities is still significant, most of them live in poverty and have the lowest indicators of human development (UNDP 2012). Therefore, bioethicists of these countries work earnestly to incorporate the topic of human rights to bioethics. Some progress has been made, for example, in Bolivia, where the majority of citizens belong to indigenous communities, "multiculturalism is pluralistic" as defined by Diaz-Couder (1998) and the various native languages and cultures are protected by Bolivian legislation and these native languages are publicly used at the same level as Spanish. Dialects, however, have been used to discriminate against native peoples in some countries such as Panama, and reinforce all other forms of discrimination. It should be noted that the linguistic definition of "dialects" for native languages is that which is locally used for private activities and excluded from public forms of communications. However, reflection about the impact of technology on the environment has opened the dialogue between Indigenous community and the dominant ones (Diaz-Couder 1998).

The Afro-American movement of Civil Rights in the United States has also influenced the Afro-American groups in the Caribbean and their claims for the right to education and political recognition. Access to universities has allowed new professionals to re-claim and re-write the historical memory of their participation in the different moments of the construction of the new independent States of Latin America or the Panama Canal. Another aspect of this educational integration is the recognition that they are as good professionals as those in the dominant groups. This recovery of their memory and self-esteem has been associated with the promotion of their culture in all aspects. The high level of education and their academic relations with the Pan-American Health Organization in the beginning of this century helped an early development of bioethics in these countries. Santo Domingo and Cuba are examples of this academic work. In The Southern Cone, the different migratory waves brought together various cultures from European and Mediterranean countries, contributing to a specific culture that is above all multicultural. The advanced level of educational and economic development of these countries has prompted the early participation of philosophers in discussions about various topics of bioethics, including recognition of the problems created by advanced technologies. Brazil is peculiar as it embraces all the cultures of America and is economically developed. Therefore, bioethicists in Brazil have risen to the challenge of considering both the problems of equity and justice, as well as the problems created by the technological advances.

The UNESCO Declaration on Race and Racial Prejudice of 27 November 1978, the UNESCO Universal Declaration on Cultural Diversity of Countries of 27 June 1989, the International Labour Organization Convention N° 169 concerning Indigenous and Tribal Peoples in Independent Countries of 27 June 1989, and the Declaration of the United Nations on the Rights of Indigenous Peoples have all contributed to the empowerment of discriminated groups to claim their rights. Gender differences represent the historical sequels to the patriarchal colonialist and indigenous societies. Although many States of Latin America participated in the meetings promoted by the United Nations in México, Cairo and Beijing and incorporated the final Declarations into their legislations, much remains to be done for the recognition of women's rights in a pluralistic society. The greatest achievements have been in education, and more women have access to the university and in some countries there are more women than men enrolled in and graduating from universities. Unfortunately women's salaries are still generally lower than men's. Health policies have always incorporated the protection of mothers and children, however, the World Health Organization (WHO) and UNICEF have recognized that there is a gap between the objectives and actual reality (UN 2015). For many societies, the concept of pluralism does not include gender as a social category different from sex, so they do not recognize the need to act upon these criteria. Likewise, homosexuality is discriminated in most countries. Finally, the dominant model is influenced by the technological revolution and the perception of its capacity to solve health and social problems. No one can ignore the benefits of technology for human health and quality of life, but academics and societies have to examine technology's negative impacts on society and health within the precautionary principle and understand its limited ability to solve social problems.

15.2 Bioethics, Latin America and UNESCO[1]

In Latin America, bioethics was introduced by Professor José Alberto Mainetti, in Argentina in the 1970s. However, only in 1994 when the Second World Congress of the International Association of Bioethics (IAB) was held in Buenos Aires, did bioethics become definitively rooted in this region. Later, in 2003, the Bioethics Network for Latin America and the Caribbean (REDBIOÉTICA) was founded in Cancun, México, in parallel with an international meeting of the Human Genome Project. The IAB World Congresses in Japan (1998) and Brazil (2002) were very instrumental in its creation. The official themes chosen for the two events "Global bioethics" and "Bioethics, power and injustice" encouraged the start of discussions relating to the search for appropriate ethical responses to moral conflicts in this region. From the beginning, it was evident that the bioethical agenda needed to be expanded beyond the biomedical and biotechnological questions. Since the outset, UNESCO has decisively supported the Network's activities and actions, initially through its Regional Office in Mexico and currently through the Social and Human Sciences Sector of the Regional Science Office for Latin America and the Caribbean, which is based in Uruguay, as well as by UNESCO's Division for Ethics in Science and Technology. It should be noted that the University of Chile, with the contribution of the Pan-American Health Organization (PAHO) and the Complutense University of Spain, had initiated bioethics education in the early 1990s with educational courses on bioethics taught in Chile, Colombia, Cuba and the Dominican Republic. The educational program was divided in two streams, one on research ethics, as a response to the research development in many countries, and the other one on clinical bioethics to respond to the difficulties of doctor-patient relationships in the new technological world. Some bioethicists organized the Latin American Federation of Bioethics Institutions (FELAIBE) to promote bioethics education through conferences (Lolas 2010). Committees on Research Ethics were formed in many countries, and some of them joined in the Forum of Latin American Research Ethics Committees (FLACEIS). The Bioethics Network, REDBIOETICA, has a portal that is currently hosted in Buenos Aires (www.redbioeticaunesco.org), and the REDBIOÉTICA UNESCO Journal, a biannual online publication, has been available on the portal since 2010 onward. This network and the UNESCO's

---

[1] Most of this part contains extracts from the presentation by Volnei Garrafa in the Open Session of the Sixteen Session of the International Bioethics Committee of UNESCO, México City, 23 November 2009, published in *Revista Redbioética/UNESCO 2010, 1(1):29–41.*

Program of Permanent Bioethics Education have developed a model of permanent education in bioethics by means of distance learning courses and offered it to health professionals, social science experts, members of ethics committees, community and political leaders and members of NGO's. Since 2006 more than 1000 participants from more than 20 countries in the region have taken these courses.

An interesting contribution that originated in Latin America is the concept of "Living Well". This is based on an ancient philosophy of life among indigenous societies of the Andean region, especially in Bolivia, and which has been incorporated into the Bolivian Constitution. In this concept, what counts is not so much wealth, i.e. the things that people produce, but rather how the things produced concretely contribute to people's lives. In formulating the "philosophy of living well" not only the material goods but also other factors are acknowledged, such as social and cultural recognition, knowledge, ethical and spiritual codes, relationships with nature, human values, and visions about the future. Within this context, the economy should be governed by living together in solidarity, without misery and without discrimination, while ensuring a minimum of necessities for everyone to survive in a dignified manner. "Living Well" expresses an affirmation of rights and social, economic and environmental guarantees. Everyone equally has the right to a decent life with assurances of health, food, clean water, pure air, adequate housing, environmental sanitation, education, work, employment, rest and leisure time, physical exercise, clothing, a retirement pension, and so on.

It is encouraging that many bioethics ideas expressed by REDBIOÉTICA in Latin America have contributed to the development of UNESCO's Universal Declaration on Bioethics and Human Rights in 2005. The IBC meeting that was held in Mexico City in November 2009 was an opportunity for Latin America to share the idea of including health and social issues within the context of the Declaration. The book entitled, *UNESCO Universal Declaration on Bioethics and Human Rights: background, principles and application* (Ten Have and Jean 2009) recounts the history of the Declaration, and REDBIOÉTICA takes the view that this is the most important collective and historical document ever constructed by Bioethics, because of its openness and significant repercussions. Article 14 of this Declaration, in particular deals with "Social Responsibility and Health", and has special value for the Network because of the Network's tireless advocacy since the start of the debates that healthcare is everyone's right and that States have a duty to provide access to health for all. This has been emphasized in the above-mentioned book in a chapter written by Martínez-Palomo, a Mexican scientist who is a member of the Board of Directors of both REDBIOÉTICA and former UNESCO's IBC member (Martínez-Palomo 2009).

Since the creation of Bioethics almost 40 years ago, there has been significant development in bioethics discourse in Latin America and the Caribbean. There is a growing trend in many countries of the region to organize National Bioethics Commissions or Boards, with the task of analyzing major moral conflicts. In universities and research centers, there are growing numbers of units or groups working on various controversial emerging topics that relate to different matters of interest to Bioethics, such as: protection of human subjects in biomedical research, especially vulnerable people; organ and tissue transplantation; new reproductive technologies; cloning; embryonic stem cells; and human genetics. In addition, several countries have already established clinical bioethics committees or healthcare bioethics committees, i.e. Chile, Argentina, Colombia, Mexico, and Uruguay. However, it should be noted that in Latin America, topics such as poverty, inequality and social exclusion are very critical bioethical issues which still need to be addressed.

The issue of "Human vulnerability" (Article 8 of the Declaration), which caused many discussions in the entire continent has created two camps among scholars. While the first group emphasizes the "protection of vulnerable people" to solve all vulnerabilities including social inequities, age or disabilities (León et al. 2013), the second group considers that protection is not enough and calls for bioethics intervention to address the external causes of human vulnerability (Garrafa et al. 2005, Nascimento and Garrafa 2010). The protective approach is based on the recognition of cultural differences, the importance of minorities, and the need for a multicultural and pluralistic education. Such protective education policies promote the recognition of differences in culture and the development of public educational policies that favor the integration and social promotion of minorities. However, the proponents of intervention believe that these policies are not enough to reduce social inequities and that the solutions lie not only in curricular changes but also in societal changes and that education has to empower students from different cultures (García et al. 2011). Social ethics seems to be the field where there is a wider gap between the understanding of vulnerability by the IBC and the development of socio-economic policies in most countries. Respect for diversity still has a long way to go if it is to mean more than a socially acceptable statement or yet another way to keep people separate by claiming that it is based on respect and not discrimination. Apart from specific topics in bioethics, Latin American bioethicists have worked on a variety of concepts which have been mentioned in the Declaration:

- Respect for cultural diversity and pluralism (Article 12) and the recognition of "indigenous and local communities" (Introduction).
- "Ethical issues related to medicine, life sciences and associated technologies as applied to human beings, taking into account their social, legal and environmental dimensions" (Article 1.1)
  - "Multidisciplinary and pluralistic dialogue about bioethical issues between all stakeholders and within society as a whole" (Article 2(e))
- "Non-discrimination and non-stigmatisation" (Article 11)
  - "Cultural diversity (Introduction)
  - "The position of women" (Introduction)
- "Equality, justice and equity" (Article 10)
  - "Equitable access to medical, scientific and technological developments as well as the greatest possible flow and the rapid sharing of knowledge concerning those developments and the sharing of benefits, with particular attention to the needs of developing countries" (Article 2(f))
  - Social responsibility "to ensure that progress in science and technology contributes to justice, equity and to the interest of humanity" (Article 14)

- "International cooperation in the field of bioethics, taking into account, in particular, the special needs of developing countries, indigenous communities and vulnerable populations" within the value of solidarity (Article 13)
- "The same high ethical standards in medicine and life science research for all human beings, without distinction"
- "Safeguard and promotion of the interests of the present and future generations" (Article 2.g).

Research ethics has probably been more influenced by the IBC's work on vulnerability than clinical ethics. This is demonstrated not only in the large number of research ethics committees in the continent, but also in the development of legislation related to research on human subjects in most countries. In many cases, such as Brazil and Uruguay among others, ethical requirements are higher than those contained in the latest versions of the Declaration of Helsinki, particularly regarding the use of placebo and continuation of treatment after the completion of clinical trials. In Brazil, for example, there is a large National Program of Control in Research Ethics coordinated by the Ministry of Health since 1996, with one National Commission and nearly 650 local ethics committees. Simultaneously, growing awareness of patient's rights has led to the development of legislation on this subject, guaranteeing not only the right to health care, but respect for patient autonomy with all its consequences, as seen in Argentina, Brazil, Chile, Panama, Uruguay. Apart from these academic debates, the political changes in many countries of Latin America such as Brazil, Bolivia, and Uruguay have favored or reinforced the empowerment and participation of different groups: women, ethnic groups, sexual minorities and others who organized themselves for the recognition of their rights to reduce the existing inequities. These groups promote the application of the values enshrined in the Convention for the Elimination of all Forms of Discrimination against Women (1981).

15.3 Comprehensiveness and Fairness in Health Services: The Impact of the UNESCO Declarations

The *Universal Declaration of Human Rights* (UN 1949), and the *Universal Declaration on Bioethics and Human Rights* (UNESCO 2005) recognize health as a right for everybody. It represents a moral obligation for the States that signed the Declarations to develop relevant public policies with an ethical approach to health promotion, prevention, treatment and rehabilitation of diseases as well as sustainable environmental development, and which policies are responsible towards present and future generations. This position meets the definition of health that takes into account the bio-psycho-social aspects (WHO, Declaration of Alma-Ata 1968; Ottawa Charter 1987). The ethical principles for decision making and resource allocation in health services were set by the World Health Organization as: efficiency (maximizing population health), fairness (minimizing health differences), utility (greatest good for the greatest number), rationality (based on epidemiological evidences and sustainable resources) and transparency (control by all organizations of society) (WHO 2011).

Most governments are applying the concepts and indicators of the Pan-American Health Organization (PAHO) to define health care programs, and prioritizing the criteria of efficiency, utility and rationality in a context of economic difficulties (see the web pages of the Ministries of Health of Colombia, Costa Rica, Mexico and Paraguay: www.minsalud.gov.co; www.ministeriodesalud.go.cr; www.salud.gob.mx; www.minsa.gob.pa). However, the problem is that most countries consider social determinants of health independently and approach each determinant individually, and ignore the success of democratic countries that create health policies with the participation of all stakeholders (PAHO 2011).

We have observed that autonomy and responsibility are frequently misinterpreted by health personnel as a disengagement from their own professional obligations and a transfer of all personal health decisions to the individual without analyzing the familiar and social local context. Health inequities are defined as avoidable and unfair differences in health due to poverty, absence of freedom, capabilities and opportunities in a specific context (Welch et al. 2012). Then what does "capabilities" mean in a context of discrimination based on poverty, gender or ethnicity? (UNDP 2012). Any discrimination contravenes Articles 10 and 11 of the *Universal Declaration on Bioethics and Human Rights*. Some of the difficulties for a full implementation of the Millennium Development Goals adopted by the UN in 2000 (Moser et al. 2005; CEPAL 2010) may be explained by these attitudes. The Goals intend to enhance human capability and wellbeing, including reducing poverty and improving health, education, gender equality, and environmental sustainability by 2015 (WHO 2008) and are thus in compliance with Article 14 of the *Universal Declaration on Bioethics and Human Rights*. Other social aspects of health such as cultural relevance at the local, national, and international levels may cause inequities in health (Marmot 2005; Lundberg and Wuemli 2012). Individualistic and consumerist models with the ensuing fragmentation of traditional community social relations in Latin America represent a collateral effect of commercial globalization. Along the same lines, some health professionals have misinterpreted individual autonomy as a total disengagement of the social context in medical interventions. As an example, medical tourism for surrogate motherhood affects poorer women and their families in countries like Guatemala, Mexico and Panama (Martinez 2011). The complexity of the relationship among health determinants obliges health professionals and stakeholders to take an inclusive approach that includes not only a guarantee of universality of access, but also the consideration of all health determinants to achieve effective action both for individual and community health. Community is understood here as a group of neighborhoods with common health determinants, or a historically cohesive group such as Indigenous and Afro-Americans, or a group historically discriminated by gender or by disease. So, bioethicists and bioethics associations in Latin America have emphasized the necessary comprehensiveness of health actions based on ethical values such as (Fortes 2000, 2005; Schramm and Kottow 2001; Tealdi 2007):

- **Integration:** means inclusiveness of everyone in health programs to achieve the principle of justice. In the context of complexity of all the mentioned factors, integration needs a trans-multi-disciplinary and pluralistic approach by health actors.[2] No doubt, the UNESCO Declarations about Human Rights, human vulnerability, discrimination and migration are important to bring the concept of integration into practice.
- **Lack of differentiation:** means non-discrimination and equity based on the universal human rights. Since the era of Hippocrates, lack of differentiation is a deontological obligation of the health professions. Equity will be reflected in the fairness of health services and health policies. Both are considered in the Universal Declaration on Bioethics and Human Rights, and non-discrimination and non-stigmatization are currently the focus of the UNESCO's International Bioethics Committee's work.
- **Community care and desegregation:** health programs like vaccination or prevention in women's health will be successful only if they are developed for and with communities. Patient care needs a comprehensive approach of all factors of health to respect the integrity of each person in his environmental context and to ensure the continuity of care. Family medicine and nursing have worked on this topic more than other disciplines because in many countries they are responsible for primary care (Brito-Silva et al. 2012; Louro Bernal 2005). The Casebook Series for the Bioethics Core Curriculum (UNESCO 2011) and the UNESCO's Gender Mainstreaming Implementation Framework (2003) were very instrumental in introducing quality of care concepts to healthcare professionals in an effort to help change attitudes.
- **Bioethics in health actions:** the difficulty results from modern medical education that divides the human body in its parts, emphasizes hospital care and segregates body, mind and society. REDBIOETHICA offers online courses on clinical and social bioethics and research bioethics that take into consideration the wholeness of bioethics (http://campus.redbioetica-edu.com.ar).
- **Integrity, honesty, soundness, and responsibility of health professionals:** are a requisite for trustworthiness of health programs and health care. From governments to each actor in health services, coherence between their actions and their declarations is necessary to maintain confidence in public policies. Doctors have to comply with their duties and they have to accept external evaluation of their work by patients, their family and society. The UNESCO Declarations have been used to reinforce these values.

---

[2] It is not enough to treat a disease; health services have to secure access to medicines; the nutritionist has to define a diet if necessary; the mental health group has to discuss the hopes, wishes and the prognosis and quality of life of the patient within the limits of his disease; the family has to help and public policies have to eliminate all discrimination. Knowledge of the importance of spirituality permits health personnel to better communicate and face the prospect death with the patient and his family. For a community, public policies have to secure access to basic needs; health promotion has to develop education programs on the ecology of diseases and to ensure empowerment of the community in health. (Pessini et al. 2009; Giovanella et al. 2012).

- **Fairness**: represents the concept of equity as the translation of equality into programs and actions. That is why an important group of Latin-American bioethicists are insisting on the need to protect vulnerable populations with interventions to ensure ethical public policies (Casado, Luna 2011). The Universal Declaration on Human Rights (1948) and the documents of UNESCO constitute an important source for understanding this concept.

15.4 Conclusion

The main theoretical and methodological tools for the development of bioethics originated in central nations and are available to researchers and practitioners throughout the world. Latin American bioethicists have also made some valuable contributions.

The first is that, as never before, the modern world has realized the need to change the old concepts in relation to the conflicts that result from development at any cost, versus sustainable development. The indigenous concept of Living Well may offer a starting point of reflection. The second is that autochthonous and peripheral cultures need to be maintained in order to respect moral pluralism and participatory democracies in the twenty-first century. Currently, only five centuries after the discovery of Latin America and the Caribbean by Europeans, their populations have the historical right to free themselves from economic, political and ethical domination,[3] always within the context of respect for universal human rights and bioethics values.

Latin American bioethics has had the objective to contribute to democracy, citizenship and human rights, via a more politicized concept of Bioethics. UNESCO has been a key supporter of this approach. The postulates of the Universal Declarations on Bioethics and Human Rights will encounter a positive echo and will have a larger impact on society and public policies when social historical factors are more favorable. Otherwise, bioethics will be limited to universities and some social groups. Bioethicists in Latin America will continue to ensure that bioethics has practical implications for its citizens.

References

Brito-Silva, K., A.F. Benjamin Bezerra, and O.Y., Tanaka. 2012. Direito à saúde e integralidade: uma discussão sobre os desafios e caminhos para sua efetivação (Right to healthcare and comprehensiveness: a discussion on the challenges and paths towards its implementation Interface). (Botucatu) vol. 16 no. 40 Botucatu Jan./Mar. 2012, Epub Apr 19, 2012. http://dx.doi.org/10.1590/S1414-32832012005000014

---

[3] Ethical domination is understood as teaching and dissemination of ethics reflections from Europe and United States with few or no references to bioethics works from other countries.

Casado, M., and F. Luna. 2011. Cuestiones de Bioética en y desde América Latina, UNESCO.
Comisión Económica para América Latina y El Caribe, CEPAL. 2010. *El Derecho a la Salud y los objetivos de desarrollo del Milenio*, Cap. V, 154. Available at: http://www.eclac.org/publicaciones/ xml/1/21541/capitulo5.pdf (in Spanish). Accessed 10 Dec 2013.
Díaz-Couder, E. 1998. *Diversidad Cultural y Educación en Iberoamérica* Revista Iberoamericana de Educación, Número 17, Mayo-Agosto, Educación, Lenguas, Culturas http://www.rieoei.org/oeivirt/rie17a01.htm. Accessed 10 Dec 2013.
Fortes, P.A.C. 2000. Bioética e saúde pública: tópicos de reflexão para a próxima década. *O mundo em saúde* 24(1): 31–38.
Fortes, P.A.C. 2005. Entre o estado, a sociedade e o indivíduo: uma reflexão bioética sobre noções divergentes de controle social e a saúde pública. *Revista Brasileira de Bioética* 1(4): 350–362.
García Castaño, F.J., R.A. Pulido Moyano, and A. Montes del Castillo. 2011. La educación multicultural y el concepto de cultura. *Revista Iberoamericana de Educación*. No. 13 – Educación Bilingüe Intercultural. http://www.rieoei.org/oeivirt/rie13a09.htm
Garrafa, V. 2009. *Redbioética* – A UNESCO initiative for Latin America and the Caribbean Conference presented in the Open Session of the Sixteen Session of the IBC – International Bioethics Committee of UNESCO. México City, 23 November 2009.
Garrafa, V., M. Kottow, and A. Saada (coordinadores). 2005. Estatuto epistemológico de la bioética Primera edición: 2005. UNESCO, Universidad Nacional Autónoma de México Impreso y hecho en México, 67–85.
Giovanella, L., O. Feo, M. Faria, and S. Tobar. 2012. Sistemas de salud en Suramérica: desafíos para la universalidad, la integridad y la equidad, 13–20. Rio de Janeiro: Instituto Suramericano de Gobierno en Salud.
León, F., R.M. Simó, L. Schmidt, and V. Anguita. 2013. Experiencias de los Comités de Ética Asistencial en España y Latinoamérica, Análisis de casos ético-clínicos. Santiago: FELAIBE.
Lolas Stepke, F. 2010. Bioética en América Latina, una década de evolución. Monografías de Actha Bioethica Nl4-2010, Universidad de Chile, Programa de Bioética OPS/OMS.
Louro Bernal, I. 2005. Modelo de salud del grupo familiar/ Model of family group health. *Rev Cubana Salud Pública* 31(4) Dec.
Lundberg, M., and A. Wuermli (eds.). 2012. Children and youth in crisis: Protecting and promoting human development in times of economic shocks. Washington, DC: World Bank. doi: 10.1596/978-0-8213-9547-9. License: Creative Commons Attribution CC BY 3.0.
Marmot, M. 2005. Social determinants of health inequalities. *Lancet* 365: 1099–1104.
Martinez, J.C. 2011. El turismo médico en Panamá. http://www.panamaqmagazine.com/2011_May/Medical_tourism_QT_2011_pg1_spanish.html
Martínez-Palomo, A. 2009. Background, principles and application A. Article 14: Social responsibility and health. In *The UNESCO Universal Declaration on Bioethics and Human Rights*, ed. H. Ten Have and M. Jean (eds.), 219–230. Paris: UNESCO Ethics Series.
Moser, K.A., D.A. Leon, and D.R. Gwatkin. 2005. How does progress towards the child mortality millennium development goal affect inequalities between the poorest and least poor? Analysis of Demographic and Health Survey data. *BMJ* 331: 1180–1182.
Nascimento, W.F., and V. Garrafa. 2010. Nuevos diálogos desafiadores desde el sur: colonialidad y Bioética de Intervención (New challenges from the south: dialogues between colonialilty and intervention bioethics). *Revista Colombiana de Bioética* 5(2): 23–37.
Pan American Health Organization/World Health Organization, Regional Office for the Americas. 2011. *Regional consultation on social determinants of health in WHO PAHO/AMRO Region*; 8–9 Aug 2011, San Jose. Available at: http://new.paho.org/cor/index.php?option=com_content&task=view&id=104&Itemid=264. Accessed 10 Jun 2013.
Pessini, L., C. de Barchifontaine, and F. Lolas Stepke (eds). 2009. *Ibero-American bioethics*. Springer.

Schramm, F.R., and M. Kottow. 2001. Principios bioéticos en salud pública: limitaciones y propuestas (Bioethical principles in public health: Limitations and proposal Cad. *Saúde Pública* 17(4):949–956

Tealdi, J.C. 2007. Bioética y Salud Pública Conferencia en Seminario Internacional "Bioética y salud pública: encuentros y tensiones", Bogotá, Universidad Nacional de Colombia, 24 de noviembre de 2006. Publicado como "Retos para la Bioética en el campo de la Salud Pública en América Latina" en Saúl Franco (ed.), Bioética y Salud Pública: Encuentros y tensiones, Bogotá, Universidad Nacional de Colombia y UNESCO, 2007, 229–243.

Ten Have, H., and M.S. Jean. 2009. *The UNESCO Universal Declaration on Bioethics and Human Rights – Background, principles and application*: Paris: UNESCO – Ethics Series.

UNESCO. 2005. Universal Declaration on Bioethics and Human Rights. www.unesco.org/bioethics/. Accessed 15 Oct 2013.

United Nations. 1948. Universal Declaration on Human Rights. 217A (iii) General Assembly Dic. 10th 1948. www.un.org/en/documents/udhr. Accessed 3 Mar 2013.

United Nations. 1981. Convention for the elimination of all forms of discrimination against women.http://www.un.org. Accessed 23 Sept 2013.

UN Millennium goals 2000. Available at: http://www.un.org/millenniumgoals/. Accessed 22 Sept 2013.

UN Millennium Development Goal 8, The global partnership for development: The challenge we face, MDG Gap Task Force report 2013.

UNDP. 2012 Report on the human development. Available at: http://undp.org

Welch, V., M. Petticrew, P. Tugwell, D. Moher, et al. 2012. PRISMA equity 2012 extension: Reporting guidelines for systematic reviews with a focus on health equity. *PLoS Medicine* 9(10), e1001333. doi:10.1371/journal.pmed.1001333.

WHO. 2011. Rio Political Declaration on Social Determinants of Health. World Conference on Social Determinants of Health. Rio de Janeiro, Brazil. 19–21 October 2011. Available: http://www.who.int/sdhconference/en/. Accessed 26 Sept 2013.

WHO. 2008. Commission on Social Determinants of Health. Closing the gap in a generation: Health equity through action on the social determinants of health: Final report of the commission on social determinants of health. Available at: Geneva www.who.org

WHO Declaration of Alma-Ata. 1968. Available at: www.who.org. (1987) *Charter of Ottawa*www.who.org

Chapter 16
Bioethics Development in Africa: The Contributions of the UNESCO International Bioethics Committee

Monique Wasunna, Aïssatou Toure, and Christine Wasunna

Abstract The International Bioethics Committee (IBC) of UNESCO was created in 1993 to provide leadership and influence the culture of bioethics in science and medicine and to ensure that human dignity, human rights and fundamental freedoms are respected. In the last two decades, the IBC has contributed immensely to the development of Bioethics in Africa by supporting the establishment of National Bioethics Committees, strengthening the capacity of these committees, training teachers in bioethics and providing ongoing direction in addressing bioethical issues in the life sciences. Africa is also represented on the membership of UNESCO-IBC and contributes to the global IBC agenda. Africa has a rich diversity of cultures and the growth of bioethics in Africa is varied as there are few trained bioethicists and few institutions of higher learning that teach bioethics. Inadequate resource mobilization to fulfill the bioethical agenda, slow progress in terms of bioethics education in Africa and the lack of a vibrant culture of bioethical discourse are among the challenges. In Africa, it is important to also engage the public in bioethical debates.

16.1 Introduction

The globalization of health research over the past 20 years has brought with it benefits such as improved medical and scientific knowledge, evidence-based policies and practices as well as increased availability and access to healthcare in many African countries. The challenges remain in the generalizability of scientific results for ethnically and genetically diverse populations; adequacy of the regulatory

M. Wasunna (✉)
Africa Regional Office, Drugs for Neglected Diseases Initiative, Nairobi, Kenya
e-mail: africa@dndi.org

A. Toure
Pasteur Institute, Dakar, Senegal

C. Wasunna
Centre for Clinical Research, Kenya Medical Research Institute (KEMRI), Nairobi, Kenya

oversight of research activities; the priority areas of health research in international, collaborative research efforts; and in the integrity of the informed consent process within varying cultural complexities (Glickman et al. 2009; Thiers et al. 2008). In response to the need for a regulatory framework in Africa and in advancing *ethos* in health research, several organizations have invested in strengthening ethics in health research across the continent. These include the United Nations Educational Scientific and Cultural Organization (UNESCO) through its International Bioethics Committee (IBC), the European and Developing Countries Clinical Trials Partnerships (EDCTP), US Fogarty International Center, UK's Wellcome Trust, Tanzanian-based African Malaria Network Trust (AMANET) and Canada's International Development Research Centre (IDRC).

The International Bioethics Committee (IBC) of UNESCO was established in 1993 to provide guidance on ethical and legal issues raised by research in the area of medicine, biological sciences and associated technologies, and to improve and reinforce knowledge in ethics (Ten Have 2006; UNESCO 2013). It should be noted that in 2002, UNESCO included ethics as a priority area for the organization. In Africa, UNESCO bioethics capacity building initiatives have supported the establishment of National Bioethics Committees, provided training of Bioethics Committees in the framework of the Assisting Bioethics Committees (ABC) and training of teachers in bioethics (Ten Have et al. 2011; Ten Have 2006).

This chapter elaborates the capacity development of research ethics committees in the last two decades and the significance of UNESCO IBC's work on the bioethics agenda in Africa.

16.2 The Landscape of Research Ethics Committees in Africa

For many years, the perception of Africa has been that of a "mystifying jungle" whose signature is anarchy, poverty, and savagery. Africa is not peculiar: like most developing countries the continent experiences a significant proportion of disease burden and preventable death yet only about 10 % of the global funding for health research is directed to addressing the continent's health problems. This global health research inequity termed the "10/90 gap" represents a fatal imbalance. The disparity in the investment in health research funding, capacity building in research and development, and the disparity in public-private partnerships has decreased in recent years with an increasing number of clinical trials conducted in Africa to develop new treatments against infectious diseases which have no territorial boundaries. Additionally, there has been a shift towards creating consortia for health research projects that are relevant to the participating communities, for example, the malaria vaccine study (Mwangoka et al. 2013) and the Human Health and Heredity in Africa Initiative. However, many countries still lack the basic infrastructure and expertise to conduct health research that meets international standards (Ali et al. 2012; Dolgin 2010; Hyder et al. 2007).

The variations in the socio-economic status, political climate, legislation, culture and history between African countries reflect the resources and infrastructure available to support health research (Hofman et al. 2013; Mathooko and Kipkemboi 2014). Despite these differences, local research ethics committees have established guidelines for the review and oversight of research in their jurisdictions in order to protect research participants from risk of harm or exploitation and to support the national health agenda. These guiding principles are adopted and domesticated from existing international documents such as the Declaration of Helsinki, *Council for International Organizations of Medical Sciences* (CIOMS), the Belmont Report and the UNESCO *Universal Declaration on Bioethics and Human Rights* (UNESCO 2005). The reality is that many countries in Africa have limited capacity to conduct effective review and oversight of health research protocols. However, more recently, targeted training programs in health research ethics and the ethics review process have been established to enhance expertise in this area (Abou-Zeid et al. 2009; Adebamowo 2007; Benatar 2007; Moodley and Rennie 2011; "SARETI" 2003). In 1993, the Council for International Organizations of Medical Sciences (CIOMS), in collaboration with WHO, provided international ethical guidelines for biomedical research involving human subjects which have been subsequently revised. The oversight and conduct of medical research in developing countries was further strengthened through the adoption and implementation of these guidelines (CIOMS 2002). Emphasis was placed on the importance of grounding biomedical research to national research and health agendas and for research to be responsive to the health priorities of participating communities (Margetts et al. 1999).

There has been a proliferation of Research Ethics Committees in Africa to support the growing number of international collaborative research projects in biomedical and social sciences, new technologies and innovations (Agnandji et al. 2012; Wasunna 2005).

However, to examine the capacity of well-established Research Ethics Committees in Kenya, Lesotho, Malawi, Mauritius, Swaziland, Seychelles, Tanzania, Uganda, Zambia and Zimbabwe, a survey was recently conducted by the East, Central and Southern Africa Health Community Secretariat in Arusha, Tanzania, using a self-assessment tool for research ethics committees (Sleem et al. 2010). The results revealed non-conformities in the minimal standards for membership, tenure and operations by some of the review boards. More strikingly, the study shows disparity in the minimum training requirements for a member to serve on the Committee (unpublished report, 2013). There are also varying degrees of stringency in the review of multi-site clinical trial protocols within and between countries (Nyika et al. 2009). While considerable time and resources must be invested to ensure an effective and efficient regulatory system, most members of ethics review boards offer their time and expertise for free by balancing their professional obligations with dedicated time to altruistically undertake ethics reviews of health research proposals.

16.3 The Impact of UNESCO-IBC's Initiatives on Africa

The following four initiatives demonstrate how instrumental UNESCO's work is in bioethics development in Africa: UNESCO Assisting Bioethics Committees Program; Regional Centers for Documentation and Research on Bioethics; Ethics Teachers Training Program; and African Representation in the Membership of UNESCO IBC.

16.3.1 UNESCO Assisting Bioethics Committees Program

The *Universal Declaration on Bioethics and Human Rights* adopted by UNESCO in 2005 advocates for the establishment of independent, multidisciplinary, and pluralist ethics committees at institutional, national, or regional levels. These committees are required to (a) assess the relevant ethical, legal, scientific, and social issues in research involving human beings, (b) provide advice on ethical problems in clinical settings, (c) assess scientific and technological developments, make recommendations, and contribute to the development of guidelines, and (d) foster debate, education and public awareness of, and engagement in, bioethics (Article 19). This guiding document which has been adopted by several African member states is intended to preserve human rights and fundamental freedoms (Langlois 2011).

Given the lack of national bioethics frameworks in many African countries, the UNESCO Assisting Bioethics Committees Project has supported the establishment of nine National Bioethics Committees (NBC) in Africa, namely; Chad, Côte d'Ivoire, Ghana, Guinea, Gabon, Kenya, Madagascar, Mali, and Togo. Discussions are ongoing to establish more of such committees in other African countries including Botswana, Malawi and Nigeria.

Kenya's National Bioethics Committee (NBC) is an example of such a committee in Africa, and we highlight the achievements of this committee as well as some of the challenges. The membership of Kenya's NBC stands at fourteen. Members are from different disciplinary backgrounds and with vast experience in serving on ethics review committees in Kenya. Among its major functions, the NBC develops guidelines and policies for ethics review, advises on specific issues of national importance and promotes ethics education and training. In 2009, just after its establishment, the NBC embarked on a training programme with UNESCO under the Assisting Bioethics Committees (ABC) Project. Since 2011, three 1-week training modules have been delivered in Kenya. During the year 2010, the Committee also developed guidelines for accreditation of institutional ethics review committees. The guidelines are now in force and have been used to accredit 19 institutional ethics committees in the country. The committee has also developed a template for material transfer agreements when biological materials are exchanged between local and foreign scientists and institutions. A draft ethical guideline about animal care and use in biomedical research is in the process of completion. It will guide institutions using animals in research and provide training for the proper care of animals. The Committee has made several landmark decisions that are often cited by review committees as well as in policy documents, such as the bio-security document and the Health Bill for Kenya being developed by the Ministry of Health and the Ministry of Science, Technology and Innovation (Simon Langat, personal communication). Additionally, the IBC has promulgated several issue-specific documents such as the Reports on the Principle of Consent (2008), the Principle of Social Responsibility and Health (2010), and the Principle of Respect for Human Vulnerability and Personal Integrity (2013). Where the national regulations are weak or non-existent, the Ethics Committees have consulted these tools to provide guidance on the responsible conduct of high-quality research relevant to the national health policy.

In Kenya, the National Bioethics Committee is planning to work on public engagement in topical bioethics issues such as initiatives to promote and implement the *Universal Declaration on Human Rights and Bioethics*. It will also engage other committees in the countries of the region to forge increased networking and collaboration in bioethics.

16.3.2 *Establishment of Regional Centers for Documentation and Research on Bioethics*

UNESCO has also established UNESCO Chairs in Bioethics in Africa. UNESCO Chairs in Bioethics have been held by representatives from Egerton University, Kenya in 1998, Côte d'Ivoire's University of Bouaké and the University of Khartoum, Sudan in 2010. Following the adoption of the *Universal Declaration on Bioethics and Human Rights* and other such declarations/conventions, bioethical discussions in science and technology is inevitable. In Kenya, the UNESCO Regional Documentation and Research Centre on Bioethics was officially inaugurated by the Director General of UNESCO on 18th May 2007 during his visit to Kenya to attend the 14th Ordinary Session of the International Bioethics Committee.

The establishment of a resource center for bioethics in the region comes at the time when Africa is trying to keep pace with new technologies and various international requirements aimed at protecting Human Rights. The Regional Bioethics Center at Egerton University, for example, has been created as a resource hub to disseminate information on bioethics to the region. The center aims to facilitate exchanges between policy makers, scholars, civil society and other interested parties on ethical, legal and social concerns stemming from advances in life sciences, especially in bioethics, of particular interest to Africa and developing countries. The centre also provides a platform for sharing information on international instrument development, challenges and priorities, and the ways and means of developing and implementing national policy frameworks in the field of bioethics. In addition, the centre promotes research and ethics education activities – both in the area of bioethics and ethics of science and technology – in particular the training and education of future scientists, policy-makers and professionals (Dr. Julius Kipkemboi of the UNESCO Regional Centre for Documentation and Research on Bioethics at Egerton University, Kenya, personal communication, 2014).

16.3.3 Ethics Teachers Training Program

Training for teachers of ethics is another contribution of UNESCO that has impacted bioethics in Africa in terms of bioethics capacity building. The overall aim of this program is to support the development of sustainable ethics education programs in Member States. In Africa, training courses have been offered in Kenya and Namibia. In Kenya, the Ethics Teacher Training Program was organized by UNESCO's Division of Human and Social Sciences and the Bioethics Chair at Egerton University in 2007. The course was attended by students from Italy, Iran, Zimbabwe, Uganda and Kenya and the course faculty came from France, Israel, The Netherlands and Kenya. It should be noted that with the help of UNESCO, regional bioethics networks have also been created in Africa. For example, the Regional Documentation and Information Centre in Egerton University, the Regional Documentation and Information Centre for Bioethics of Science and Technology in the Academy of Scientific Research and Technologies in Cairo and the Bioethics Network on Women's Issues in the Arab Region in Egypt are some of the existing bioethics networks.

16.3.4 African Representation in the Membership of UNESCO IBC

Since the IBC's inception, African countries have played a major role as members of the IBC. There are currently five members from Africa on the committee. Some of the African members of the IBC are also members of their respective national bioethics committees. This means that progressively over the years, African countries have built capacity in bioethics and bioethical reflections on major issues affecting their countries. It should be noted that such membership has created a platform for experts from Africa not only to gain from but also to contribute to the global bioethics discussion.

16.4 Current Challenges

Despite the contributions highlighted above, there are a number of challenges worth noting. One major challenge is the slow pace in the development of bioethics education on the continent. Currently, very few African countries offer bioethics training programs at their various universities. With the increasing globalization of biomedical research and the growing interest in the field of bioethics on the continent, African Member States will need to take bioethics education seriously. It is important that the 'training the teachers in bioethics' program be sustained to provide a critical mass of bioethics teachers who will in turn develop context-specific bioethics modules for their countries. This will however require support from all relevant stakeholders, such as the ministries of education, universities and policy makers to sustain this initiative.

Another challenge is the limitation caused by a language barrier. For instance, the training of teachers program has a limited impact because courses are usually offered in English, thus only English speaking countries in Africa can benefit. Francophone countries in particular do not have teaching materials in French and cannot access the existing frameworks for training due to the language barrier. This is important as the system of training in bioethics is virtually non-existent in non-English speaking academic institutions and there must be continuing reflection and work to introduce bioethics into the curricula of the Francophone countries of Africa.

A major challenge in the field of bioethics is ensuring the support of national consciousness and political commitment. African countries should be encouraged to continue establishing independent, multidisciplinary national bioethics committees and to empower these committees to effectively engage in ethical debates raised by inequitable access to health care, new biotechnologies, and advances in science. This, however, also cannot be achieved without political will and the involvement of policy makers. Therefore, policy makers will also need to be educated about the important role of bioethics in national legislatures.

16.5 Conclusion

In the last 20 years, the UNESCO IBC has contributed positively to the development of bioethics in Africa. The establishment and capacity building of national bioethics committees has borne fruit. Trained bioethicists can now provide context-specific guidance, build capacity for ethics review, and improve the quality of ethical health research studies. African scientists, bioethicists, civil societies, advocacy groups and professional societies must actively contribute to the international ethics debates through dialogue and exchange of experiences to bring about equality in the global research agenda and research initiatives. Best practices and reflections in bioethics should be harnessed and promoted without infringing on the rights of individuals and communities, even when confronted by emerging technologies and globalization. The extent of bioethics education in Africa is unknown and its current status is also difficult to ascertain. There are still many challenges in bioethics education. It requires immense resources and goodwill for it to develop to a significant level. Deliberate efforts are needed to increase bioethics education in Africa and embrace it in schools, universities and other public and private institutions.

References

Abou-Zeid, A., M. Afzal, and H.J. Silverman. 2009. Capacity mapping of national ethics committees in the Eastern Mediterranean Region. *BMC Medical Ethics* 10: 8.

Adebamowo, C.A. 2007. West African bioethics training program: Raison D'être. *African Journal of Medicine and Medical Sciences* 36(Suppl): 35–38. Retrieved from http://www.ncbi.nlm.nih.gov/pubmed/17703562?dopt=abstract.

Agnandji, S.T., V. Tsassa, C. Conzelmann, et al. 2012. Patterns of biomedical science production in a sub-Saharan Research Center. *BMC Medical Ethics*. doi:10.1186/1472-6939-13-3.

Ali, J., A.a. Hyder, and N.E. Kass. 2012. Research ethics capacity development in Africa: Exploring a model for individual success. *Developing World Bioethics* 12(2): 55–62. doi:10.1111/j.1471-8847.2012.00331.x.

Benatar, S. 2007. Research ethics committees in Africa: Building capacity. *PLoS Medicine* 4(3), e135.

Council for International Organizations of Medical Sciences. 2002. *International* ethical guidelines for biomedical research involving human subjects. *Bulletin of medical ethics*, 17–23. Retrieved from http://www.ncbi.nlm.nih.gov/pubmed/14983848

Dolgin, E. 2010. African networks launch to boost clinical trial capacity. *Nature Medicine* 16(1): 8. doi:10.1038/nm0110-8a.

Glickman, S.W., J.G. McHutchison, E.D. Peterson, et al. 2009. Ethical and scientific implications of the globalization of clinical research. *The New England Journal of Medicine* 360(8): 816–823. doi:10.1056/NEJMsb0803929.

Hofman, K., Y. Blomstedt, S. Addei, et al. 2013. Addressing research capacity for health equity and the social determinants of health in three African countries: The INTREC programme. *Global Health Action* 6(April 2013), 19668. Retrieved from http://www.pubmedcentral.nih.gov/articlerender.fcgi?artid=3617877&tool=pmcentrez&rendertype=abstract

Hyder, A.A., R.A. Harrison, N. Kass, and S. Maman. 2007. A case study of research ethics capacity development in Africa. *Academic Medicine: Journal of the Association of American Medical Colleges* 82(7): 675–683.

Langlois, A. 2011. The global governance of bioethics: Negotiating UNESCO's Universal Declaration on Bioethics and Human Rights (2005). *Global Health Governance* V(1–23).

Margetts, B., L. Arab, M. Nelson, and F. Kok. 1999. Who or what sets the international agenda for research and public health action? *Public Health Nutrition* 2(3): 235–236.

Mathooko, J., and J. Kipkemboi. 2014. African perspectives. In *Handbook of global bioethics*, ed. H.A.M.J. ten Have and B. Gordijn, 253–268. Dordrecht: Springer.

Moodley, K., and S. Rennie. 2011. Advancing research ethics training in Southern Africa (ARESA). *South African Journal of Bioethics and Law* 4(2): 104–105.

Mwangoka, G., B. Ogutu, B. Msambichaka, et al. 2013. Experience and challenges from clinical trials with malaria vaccines in Africa. *Malaria Journal* 12(1): 86. doi:10.1186/1475-2875-12-86.

Nyika, A., W. Kilama, R. Chilengi, et al. 2009. Composition, training needs and independence of ethics review committees across Africa: Are the gate-keepers rising to the emerging challenges? *Journal of Medical Ethics* 35(3): 189–193. Retrieved from http://www.pubmedcentral.nih.gov/articlerender.fcgi?artid=2643018&tool=pmcentrez&rendertype=abstract

SARETI. 2003. *South African research ethics training initiative: Advanced training in health research ethics*. Retrieved March 30, 2014, from http://sareti.ukzn.ac.za/Homepage.aspx

Sleem, H., R.A. Abdelhai, I. Al-Abdallat, et al. 2010. Development of an accessible self-assessment tool for research ethics committees in developing countries. *Journal of Empirical Research on Human Research Ethics* 5(3): 85–96. quiz 97–98.

Ten Have, H. 2006. The activities of UNESCO in the area of ethics. *Kennedy Institute of Ethics Journal* 16(4): 333–351.

Ten Have, H., C. Dikenou, and D. Feinholz. 2011. Assisting countries in establishing national bioethics committees: UNESCO's Assisting Bioethics Committees project. *Cambridge Quarterly of Healthcare Ethics: CQ: The International Journal of Healthcare Ethics Committees* 20(3): 380–388. doi:10.1017/S0963180111000065.

Thiers, F.A., A.J. Sinskey, and E.R. Berndt. 2008. Trends in the globalization of clinical trials. *Nature Reviews Drug Discovery* 7(1): 13–14. doi:10.1038/nrd2441.

UNESCO. 2005. UNESCO Declaration on Bioethics and Human Rights. Available at: http://www.unesco.org/new/en/social-and-human-sciences/themes/bioethics/bioethics-and-human-rights/. Last visited 23 May 2014.

UNESCO. 2013. 19932013: 20 years of bioethics at UNESCO.

Wasunna, A. 2005. The development of bioethics in Africa. In *Bioetica ou Bioeticas na Evolicao das Sociedas, Coimbra, Grafica de Coimbra*, 331–334.

Chapter 17
Bioethics in East Asia: Development and Issues

Myongsei Sohn

Abstract The East Asian countries have a long history of cultural achievement with their own ethical perspectives rooted in a rich religious and philosophical background. These traditions acknowledge the importance of family in one's existence and, as compared to the West, there is less emphasis on individuals. During the rapid transition to modern Western medicine, these traditions were not considered in the development of local bioethical frameworks in East Asia. At the same time, East Asian nations were experiencing changes in social and economic structures, mostly led by developments in science and technology. Thus, East Asian societies were initially quite unfamiliar with the newly introduced concept called "bioethics". To deal with sensitive bioethical issues the Western bioethical approach was adopted, without sufficiently taking into account local traditions. This chapter presents a historical development of bioethics followed by a description of the main issues and challenges in the development of bioethics discourse in East Asian countries.

17.1 Introduction: Global and Local Bioethics

The nations in East Asia – China, India, Japan, Korea, Singapore and Thailand- have a long history and cultural achievement with their own ethical perspectives rooted in rich religious and philosophical traditions. *Carata Samhita* of India and *Lun ta i ching-ch'eng*' of traditional Oriental medicine are examples (Unschuld 1979). These traditions are committed to advancing human welfare, acknowledging the importance of family in one's existence, and with less emphasis on individualism. During the rapid transition to modern Western medicine, these traditions were not referred to in the development of local bioethical frameworks in East Asia. At the same time, these East Asian nations were experiencing changes in social and economic structures, mostly led by development in science and technology. Thus, East Asian societies were initially quite unfamiliar with the newly introduced

M. Sohn (✉)
Department of Medical Law and Ethics, College of Medicine, Yonsei University,
Seoul, South Korea
e-mail: msohn53@gmail.com

concept called "bioethics". To deal with sensitive bioethical issues, the Western bioethical approach was adopted without sufficiently taking local tradition into account. Increasingly, it has been recognized that there is a need to ensure a balance between the benefits and harms of scientific knowledge and technology. Researchers from East Asian countries are the most active participants in world-wide bio-medical scientific competitions, and many research studies are undertaken in East Asia whether by local researchers or in collaboration with Western researchers. With increasing participation in these research activities, researchers have been confronted with numerous ethical issues and were themselves instrumental in calling for the need of bioethical governance over scientific research. Furthermore, as most biomedical research studies became more global in scale, researchers and regulators were required to comply with a variety of international research norms developed by international bodies such as the World Health Organization and the UNESCO International Bioethics Committee (IBC).

As emphasized by the UNESCO International Bioethics Committee (IBC), it is important to ensure that bioethical regulations and declarations are locally relevant (Bagheri 2011), and research communities in East Asia have realized that global ethical norms must be adapted by considering local norms to create pertinent and practicable rules for the particular community.

17.2 The Development of Bioethics in East Asia

Bioethics is not a single academic discipline; rather it is a field of interdisciplinary activities including philosophical exploration, empirical studies, legal studies, social activism, and policy development. Some of the early pioneers in East Asian bioethics were medical educators who tried to strengthen the professionalism in medicine. There are various reasons to introduce bioethics discourse to a country: sheer academic interest in applied ethics; help inform a policy before investment in Research and Development; or introduce a topic of bioethics in which society should become engaged. Some cases related to research fraud compelled the development of bioethics, such as WS Hwang's Nuclear Transfer Stem Cell scandal in South Korea (NTSC scandal 2005). These cases caused public alarm and first elicited professional responses in positive as well as negative ways, and then garnered academic interest. Bioethicists from various disciplinary backgrounds came together, trying to understand and resolve the concerns raised by each case. Meanwhile ethical frameworks began to take shape, depending on their urgency and potential (Sulmasy 2010). Following the introduction of bioethics into East Asian countries, bioethics analysis was initially focused more on the legalization of practices involving bioethical issues such as abortion, organ transplantation, genomic research and cloning. Most ethical issues were discussed in response to the regulatory measures proposed for certain practices but were also related to clinical and research activities. In the case of genetic research and human cloning, robust ethical debates took place involving various stakeholders, including the public, and consequently many governments enacted human subject protection codes and required national or institutional research ethics oversight. Such trends in enhancing governance of biomedical research continue and many countries have adopted the necessary legislations to regulate controversial areas of clinical practice.

As observed by several commentators in this region, bioethics was first imported, and then developed (Akabayashi 2009; Li and Cong 2008). Although it can be said that bioethics standards were initially adopted to address pressing administrative and practice issues, there has now been time for reflection on how these Western bioethical approaches can be adapted to the local context. While ethical principles must be accepted because they are based on a common understanding, the development of bioethics and bioethical approaches still require the on-going processes of: introduction; execution; critical appraisal of the theory; and the adoption of bioethical theories, frameworks and approaches. For example, countries have chosen one of three approaches to research ethics oversight: autonomous regulation by professional associations; the establishment of institutional ethical oversight mainly by research ethics committees, also called Institutional Review Board (IRB); or national bioethics advisory bodies established by bioethics-influenced legislation. It is worth noting that bioethics discourse in East Asian countries developed as a result of the interaction between government, researchers, physicians and bioethicists. Government has been a supporter of bioethics by funding activities and enacting legislation, even though bioethicists are not always in agreement with the government's approach. While limited in their ability to effect change, researchers and bioethicists are sensitive to the ethical issues raised by research and health care, and have launched discussions on many important bioethical issues. It should be noted that the UNESCO-IBC's documents on bioethical issues have also provided support to the relatively few bioethicists when engaging the public's interest on bioethical issues. In addition, the UNESCO-IBC has responded to bioethical issues with deep insight, developed guidelines by international consensus and also suggested practicable alternatives.

The above summary has described the social and political influences on the development of East Asian ethical approaches. Next, the influences of law, institutional and organizational policy and academia on current East Asian bioethics will be discussed.

17.3 Bioethics and Legislative Activities

Bioethics was introduced around the 1970s, yet many countries had already identified ethical issues and enacted laws to regulate related practices, for example abortion laws. Since that time, many more bioethics-related laws have been enacted on a variety of topics, such as organ transplantation, medical practice (professionalism, informed consent), end-of-life care, biomedical research, human research subject protection, and human reproduction. Many of these laws reflect each country's particular understanding of socio-cultural norms and human rights principles. However, laws related to international bioethical issues, such as organ trafficking, stem cell research, medical tourism, and multinational clinical trials, remain greatly influenced by international regulations and guidelines. For instance, while international biomedical research is directed by the Declaration of Helsinki and CIOMS guidelines in general, genetics and human cloning are guided by UNESCO Declarations (WMA 2008; CIOMS 2002; UNESCO 1997).

As the legislative process in a given parliament is a political response to social attitudes and reflects the necessities of that time, a review of legislative history tells us something about a particular society's ethical norms. For example, Korean laws have resulted from the differing social dynamics: the abortion act was set up during a period of rapid economic development (1960 to 1970s); the organ transplantation act was introduced after rapid development in clinical medicine in 1990s; and Bioethics and Safety Act resulted from turbulent stem cell and human cloning debates (Hahm and Lee 2012). In the year 1999, the Korean National Commission for UNESCO (KNCU) organized the 'Consensus Conference on Cloning' in order to listen to the public's opinion on human cloning, and through which the citizens reached consensus that: human life starts at the point in time right after fertilization, and human or embryo cloning should be prohibited because human dignity would be harmed if embryonic tissue were used as research material. This consensus was the starting point of Korea's Bioethics and Safety Act (2005).

In 2002, the KNCU organized a workshop on 'Science Technology and Ethics' celebrating the World Science Day for Peace and Development, in which participants opened up discussions on topics such as human cloning and lack of privacy of personal information when using the Internet (KNCU 2002).

17.4 Bioethics Infrastructure: Research Ethics Committees

Bioethics committees and research ethics committees are important contributors to bioethical deliberations and decision-making processes (UNESCO 2010). In many East Asian countries, bioethics development was initiated by discussion about research ethics and strengthened by the establishment of research ethics committees (RECs). All countries in East Asia except Thailand, have established both committees for the review of research studies as well as national bioethics (advisory) committees, many of which were set up after the successful cloning of Dolly, the sheep in 1996.

In East Asian countries, ethical oversight of research and consultation is a basic role of REC. These committees were introduced in the late 1980s to early 1990s, and research ethics review became mandatory in the early 2000s. The current issue is how to improve the quality of ethics review by these committees. Domestic accreditation systems (i.e. the Korean IRB registration system) and international accreditation organizations such as The Forum for Ethical Review Committees in Asia and the Western Pacific Region (FERCAP 2011) are currently operational (Son et al. 2014). Even though accreditation or certification of ethics review committees is not required, all committees have developed and follow Standard Operating Procedures (SOPs), and regularly offer continuing education to IRB members to ensure quality ethics review.

Bioethics advisory committees are established in governmental structures or work as independent non-governmental organizations (WHO 2014). In some countries the National Bioethics Committees are established within the existing governmental structure, for instance, in China by the Ministry of Health, and in India by the Ministry of Science and Technology. These committees are assigned to act as an independent body for policy advice and counseling on ethical issues as they emerge in clinical, research and public health practice. The UNESCO IBC's guidelines also provide practical information not only on the structure of a National Bioethics Committee, but also on its composition, work, policies, education, etc. (UNESCO 2005). It should be noted that national bioethics committees do not have a unified structure, for instance in some countries national bioethics committees function as arm's length advisory bodies independent from the government. Japan has a unique configuration somewhat in between the two described above and has two bioethics advisory structures: one is the "Expert Panel of Bioethics" established in the Council for Science and Technology Policy which operates as the national bioethics committee; and the other is a series of ethics committees constituted within relevant ministries, such as the Ministry of Education and the Ministry of Health. In either form, ethics committees can be influential in beginning and sustaining discussions as well developing bioethics policies, standards or guidelines on important topics.

The work of the UNESCO IBC has also assisted East Asian ERCs in developing guidelines for sensitive issues in biomedical research involving human subjects, for example, obtaining informed consent in biomedical research. There are two articles from the *Universal Declaration on Bioethics and Human Rights* that are instrumental in dealing with this issue: Articles 6 and 7 address the principle of consent itself and the situation where the person does not have the capacity to consent (UNESCO 2005). Moreover, the importance of informed consent has been highlighted in two publications by the UNESCO-IBC (INCU 2003; UNESCO 2008). Recognizing the necessity for an active discussion among experts regarding the increasing focus on informed consent in South Korea, the KNCU organized a session entitled "Consent in Medicine and Life Science Research" in 2007.

It should be noted that while the role of RECs is indispensable and research ethics review is accepted as necessary before the commencement of biomedical research, the practice of clinical ethics consultation is less common in East Asian countries. Singapore has a well-organized clinical ethics consultation service (Singapore Ministry of Health 2014), while other countries have little recognized clinical ethics consultation services in their healthcare institutions. This may be due to the paternalistic attitude of medical professionals and patient's family members, or due to other factors related to medical practice. As the issue of end-of-life care becomes the focus of social debate, the importance and the role of clinical ethics consultation will hopefully become better recognized and clinical ethics committees will become an important resource for providing consultation services in healthcare institutions.

17.5 Bioethics Education and Academic Development

Bioethics has a long and rich history in academic circles and early bioethicists were scholars in philosophy, law and medical science. Bioethicists in East Asia make an effort to integrate their traditional thinking into bioethical reasoning by exploring Confucianism, Hinduism, Buddhism and Islamic traditions. Accordingly, in deliberating the ethics of cutting edge biosciences, bioethicists try to justify their approach by using both traditional reasoning and bioethical reasoning. However, as bioethics emphasizes a realistic understanding of the topic at hand, empirical research in bioethical issues is increasing. In East Asia, empirical research in ethics initially began with anthropological, comparative studies; however, feminist, narrative and historical approaches are now also integrated into research methodology. The increasing interest in bioethical discussion has also been instrumental in the creation of academic bioethics associations in many East countries, which provide a forum for active participation in bioethical discussions in the region.

Bioethics also is a major focus in professional education as increasing numbers of medical professionals and scientists are taking bioethics courses (MOST 1998; Sohn 1998). These courses are either included in professional curricula, or held during professional conferences and meetings. In some cases relevant institutions within governments conduct bioethics courses for professionals. The content of courses includes the following topics: integrity research activities; clinical etiquette; distributive justice in resource allocation; informed consent; confidentiality; and end-of-life issues. Other topics include how to teach ethics as well as education about policy development and legislation. These courses are currently important topics for professional education because professionals are expected to seamlessly interpret and fulfill current law, policy and ethics standards.

In terms of capacity building in bioethics education, the UNESCO bioethics education programs as well as its Declarations have greatly contributed to the development of bioethics in the region in many ways. For instance in 2009, with the help of UNESCO, an International Symposium on Building and Operating the Ethics Committee was organized in Shanghai with more than 400 participants from all over China. Participants discussed the following subjects: (1) Establishment and Evaluation of the Ethics Review Committee, (2) the SOPs of the Ethics Review Committee, (3) The Management and Supervision of the Ethics Committee, and (4) The Continuing Education Program for the Members of Ethics Review Committees. Since then, annual bioethics conferences or training programs were conducted in Shanghai, Beijing and many other cities in China. These events were organized by the National Ethics Committee, the Chinese Ethics Society and Bioethics Committee of the Chinese Society for Philosophy of Nature, Science and Technology.

17.6 Important Bioethical Issues

17.6.1 Reproduction

The issue of abortion is among the oldest, but most hotly debated, ethics topics in East Asia. Each country has a unique and fascinating cultural and religious background, and tries to balance the unyielding, often opposing perspectives on the sanctity of life and reproductive autonomy. Like other bioethical issues, but more acutely with reproduction, it is complicated with social-economical-political conditions.

Most countries have laws regulating abortion and artificial reproduction practices. However, the permissibility, and the scope and conditions of legal abortion remain controversial. Abortion, especially illegal abortion related with sex selection has been problematic. For example, India responded to this problem with legislation to prohibit the practice of sex selection by enacting the Pre-conception and Prenatal Diagnostic Techniques (Prohibition of sex selection) Act 1994, amended in 2003 (Ministry of Health of India 2003). Korea recently revised the prohibition to notify the parents of the fetus' gender because of a successful constitutional challenge. The 'prohibition to notify' clause was introduced in 1987 to prevent abortion performed for the purpose of sex selection, but the Constitutional Court of Korea ruled this provision as unconstitutional (Yang 2009).

17.6.2 Organ and Cell Transplantation

Organ trafficking and transplantation tourism have become chronic ethical problems in East Asia (Bagheri 2007). As the unethical practice of organ selling increases, each country has adopted measures to prevent organ sales and to protect recipients from unsafe organs. There are also concerns arising out of recent advances in the field of stem cell technology: many potential recipients cross international borders with the hope of a cure by cell transplantation, whether adult stem cell or embryonic stem cell transplantation. The problem still remains unresolved in East Asian countries although many have adopted and endorsed the guidelines developed by international organizations such as, the World Health Organization's *Guiding Principles on Human Cell, Tissue and Organ Transplantation* (WHO 2010); UNESCO-IBC's *Report on the Principle of Non-discrimination and Non-stigmatization* (2014); the *Recommendation of the Asian Task Force on Organ Trafficking* (2008); and the *Istanbul Declaration on Organ Trafficking and Transplant Tourism* (2008).

17.6.3 Genomic Research

In the late 1990s, there was a succession of ground-breaking biomedical scientific discoveries, for example: stem cell therapy; completion of the human genome; and xeno-transplantation. Consequently, many governments such as South Korea have increased their research and development budget for human genetic and genomic research thus helping the growth of their technical competitiveness to international levels (BPRC 2011b). Many countries carefully considered the UNESCO-IBC's documents, especially the IBC's *Report on Food Plant Biotechnology and Ethics* (1996) and the *Universal Declaration on the Human Genome and Human Rights* (1997), as they developed laws. Genomic research had been perceived as quite important in the field of science because it has caught the public's attention.

In recent years, the advancement of life science technology is largely focused on the analysis of genetic information and arguments that this data should be widely shared and used, without limitations. However, a person's genetic information contains sensitive personal information that individuals may not know, or not be willing to disclose to others (such as employers, insurers, family, etc.). Individuals are now confronted with significant privacy issues that must be handled with care. Some countries have enacted legislation to protect the privacy of individuals. Korea, for example, has ensured privacy protection in genetic or genomic research through its Bioethics and Safety Act. In India, the National Bioethics Committee took steps to protect privacy with the drafting of its "Ethical Policies on the Human Genome, Genetic Research and Services" by taking special note of the *UNESCO Universal Declaration on the Human Genome and Human Rights* (1997) as well as the Guidelines issued by the Indian Council of Medical Research (2000).

Yet, ethical solutions must continually be re-evaluated and re-imagined to respond to the rapid development of technology. For example, the availability of large data stored on cloud technology for public use challenges current conceptions of privacy and the protection of a person's genetic information.

17.6.4 End-of-Life Issues

Healthcare is a human rights issue, and each government is charged with providing the best affordable care to its citizens. Meanwhile, the ideas of dying well or the right to a good death are newly introduced in many societies. Despite these goals, however, in many East Asian countries euthanasia is strictly prohibited and is considered equivalent to a criminal act of murder. Several legal cases in Korea and Japan have highlighted the controversial issues of euthanasia and withdrawing life-sustaining treatment (Lee 2011). As the legal process had been delayed, public activism filled the void (Kim et al. 2010; BPRC 2011a). Given the intense controversy, both governments were reluctant to adopt a position on death with dignity:

leaving most hospitals wondering how to interpret their patients' advance directives, and how these directives influence standard practice. In Korea, the Ministry of Health and Welfare is currently in the process of developing guidelines for advance directives, and it is anticipated they will help clarify the most important values in these complex ethical debates (MOHW and AIBHL 2013). Other nations, such as Singapore and Taiwan, have "death with dignity" laws, which allow a capable person to request an assisted death after a detailed consultation and documentation process.

East Asian countries are experiencing difficulties in ensuring that the patient's own choices in end-of-life care planning are respected because surrogate decision making is common, and sometimes family members strongly argue that they should make decisions on behalf of the patient to protect the patient from psychological or emotional harm. The debate continues, but the importance of understanding patient's own wishes is recognized and slowly the situation will be changed.

17.7 Conclusion

Despite the many struggles that East Asian countries have faced, they have succeeded in building capacity to respond to bioethical issues newly raised with the advancement of biomedicine and biotechnology. As in other regions, the UNESCO IBC, through its bioethics programs and Declarations, has influenced bioethics development and capacity building in this region as well. The dedicated membership of bioethics experts from East Asia in the UNESCO IBC has also paved the way for mutually beneficial collaborations with bioethics experts from other regions. East Asian countries are now experienced in scientific technology and bioethics, and are capable of influencing standard setting in the international community while at the same time sharing their own knowledge and culture with others. The current state of bioethics knowledge, capacity and expertise of these countries is an example of the successes which can be realized through international support and collaboration, and is certainly proof that sincerity and a helping hand can make a difference. There still remains much work to be done to address and explore bioethical issues such as traditional medicine, relational ethics, nano-ethics, and other issues of medical practice. In addressing these issues, the role of the UNESCO IBC will inevitably become even more central. East Asian countries' contribution to international dialogue and cooperation for ethical practice becomes more and more necessary.

Acknowledgement The Author deeply appreciates Qingli Hu (China), H. Sharat Chandra and P. N. Tandon (India) for the information of national bioethics development and issues in their countries.

References

Akabayashi, A. 2009. Bioethics in Japan, 1980–2009: Importation, development, and the future. *Asian Bioethics Review* 1(3): 267–278.

Bagheri, A. 2007. Asia in the spotlight of international organ trade: Time to take action. *Asian Journal of WTO and International Health Law and Policy* 2(1): 11–24.

Bagheri, A. 2011. The impact of the UNESCO Declaration in Asian and Global Bioethics. *Asian Bioethics Review* 3(2): 52–64.

Bioethics Policy Research Center (BPRC). 2011a. *Discontinuation of life-sustaining treatment and advance directives*, 110. Seoul: Bioethics Policy Research Center.

Biotech Policy Research Center (BPRC). 2011b. *Functional analysis of human genome and its application*, 118. Biotech Policy Research Center.

Council for International Organizations of Medical Sciences. 2002. International ethical guidelines for biomedical research involving human subjects. Available at: http://www.cioms.ch/publications/guidelines/guidelines_nov_2002_blurb.htm

Hahm, K., and I. Lee. 2012. Biomedical ethics policy in Korea: Characteristics and historical development. *Korean Journal of Korean Medical Science* 27(Suppl): 76–81.

Indian Council of Medical Research. 2000. Ethical policies on the human genome, genetic research and services. Available at: (http://dbtindia.nic.in/uniquepage.asp?ID_PK=113

Israel National Commission for UNESCO. 2003. *Informed consent*. Israel National Commission for UNESCO.

Istanbul Declaration on Organ Trafficking and Transplant Tourism. 2008. Steering Committee of the Istanbul Summit. Organ trafficking and transplant tourism and commercialism: The Declaration of Istanbul. *Lancet* 372: 5–6.

Kim, S., et al. 2010. A Korean perspective on developing a global policy for advance directives. *Bioethics* 24(3): 113–117.

Korean National Commission for UNESCO (KNCU). 1999. Citizen panel report: Consensus Conference on cloning. Seoul: Korean National Commission for UNESCO.

Lee, E.-Y. 2011. A shared decision making model in the end-of-life decisions and the relational autonomy. *Medical Law and Ethics*. Seoul: Yonsei University, 104.

Li, H., and Y. Cong. 2008. The development and perspectives of Chinese bioethics. *Journal International De Bioethique* 19(4): 4–12.

Ministry of Health and Welfare Korea, Asian Institute for Bioethics and Health Law. 2013. Developing medical decision-making process in the end-of-life care. Seoul.

Ministry of Health of India. 2003. Pre-conception and Prenatal Diagnostic Techniques (Prohibition of sex selection) Act 1994 amended in 2003. Available at: http://pbhealth.gov.in/pdf/PC%20&%20PNDT%20Rules%201994/PC%20&%20PNDT%20Rules%201994.pdf. Last visited 30 Jun 2014.

Ministry of Science and Technology (MOST). 1998. Study on the present status of biotechnology and bioethics and Korea's response to them. Ministry of Science and Technology. Gwacheon: 241.

Singapore Ministry of Health. 2014. National Medical Ethics Committee guidelines. http://www.moh.gov.sg/content/moh_web/home/Publications/guidelines/national_medical_ethics_committee_guidelines.html. Accessed 7 Jul 2014.

Sohn, M. 1998. Work and prospect of education on ethics in medical colleges – In view of educational experience of Yonsei University College of Medicine. *Korean Journal of Medical Ethics* 1(1): 14.

Son, S., S. Hong, and Y. Kim. 2014. Registration status of institutional review board in Korea, 2013. *Public Health Weekly Report*, KCDC. Available at: http://www.cdc.go.kr/CDC/info/CdcKrInfo0301.jsp?menuIds=HOME001-MNU1132-MNU1138-MNU0037-MNU1380 (Korean).

Sulmasy, D.P. 2010. Research in medical ethics: Scholarship in "substituted judgment". In *Methods in medical ethics*, 2nd ed, ed. J. Sugarman and D.P. Sulmasy, 295–314. Washington, DC: Georgetown University Press.

The Forum for Ethical Review Committees in the Asian and Western Pacific Region. 2011. Available at: http://www.fercap-sidcer.org/publications.php. Last visited 10 Jul 2014.

UNESCO. 1996. Report on food, plant biotechnology and ethics. Available at: http://portal.unesco.org/shs/en/files/2299/10596448761FoodCIB4_en.pdf/FoodCIB4_en.pdf. Last visited 28 Aug 2015.

UNESCO. 1997. UNESCO Universal Declaration on the Human Genome and Human Rights. Available at: http://portal.unesco.org/en/ev.php-URL_ID=13177&URL_DO=DO_TOPIC&URL_SECTION=201.html

UNESCO. 2005. Guide No. 1 Establishing Bioethics Committees. Available at: http://unesdoc.unesco.org/images/0013/001393/139309e.pdf. Last visited 10 Jul 2014.

UNESCO. 2010. *National bioethics committees in action*, 9–11. Paris: UNESCO.

UNESCO. 2014. Report of the IBC on the principle of non-discrimination and non-stigmatization. Available at: http://unesdoc.unesco.org/images/0022/002211/221196E.pdf. Last visited 10 Jul 2014.

UNESCO International Bioethics Committee (IBC). 2005. Universal Declaration on Bioethics and Human Rights. Paris.

UNESCO International Bioethics Committee (IBC). 2008. Report of the IBC on Consent. Paris: UNESCO International Bioethics Committee: 57.

Unschuld, Paul U. 1979. *Medical ethics in imperial China: A study in historical anthropology*, 26. California: University of California Press.

World Health Organization. 2010. WHO guiding principles on human cell, tissue and organ transplantation. Available at: http://www.who.int/transplantation/Guiding_Principles Transplantation_WHA63.22en.pdf?ua=1. Last visited 28 Aug 2015.

WHO. 2014. National ethics committees in the South-East Asia region. http://www.who.int/ethics/committees/searo/en/. Last visited 10 Jul 2014.

World Medical Association (WMA). 2008. WMA Declaration of Helsinki- Ethical Principles for Medical Research Involving Human Subjects. Available at: http://www.wma.net/en/30publications/10policies/b3/

Yang, H. 2009. A social-scientific perspective on the constitutionality of an article in the medical law regarding the prohibition of notifying the sex of the fetus. *Seoul Law Journal* 50(4).

Index

A
Abortion, 155, 186–188, 191
Acupuncture, 75
Africa, 7, 26, 27, 44, 105, 130, 176–181
Altruism, 27
Applied ethics, 3, 18, 159, 186
Arab region, 151–160, 180
Argentina, 105, 166, 168, 169
Asian Task Force, 93, 191
Assisting Bioethics Committees Program, 24, 130, 142, 146, 159, 176, 178–179
Australia, 26, 49, 88
Austria, 130, 140
Autonomous decision, 50–52, 111
Autonomy, 3, 4, 30, 31, 34, 36, 38, 40, 44–53, 76–78, 110–111, 137, 153–155, 159, 164, 169, 170, 191
Avicenna, 152, 156
Azerbaijan, 142, 145–147

B
Bahrain, 151
Beijing Declaration, 76
Belaru, 145, 146
Belgium, 140
Bellagio Task Force, 98
Belmont Report, 3, 35, 128, 177
Beneficence, 3, 35, 67, 76, 154, 157, 164
Benefit sharing, 9, 33, 63–68, 76, 117
Biobank, 81–88, 94, 141
Biobank legislation, 85
Biodiversity, 31, 36, 58, 68, 69, 78, 104
Bioethical challenge, 127, 155
Bioethical governance, 186

Bioethical principles, 6, 76, 160
Bioethical reasoning, 190
Bioethical responsibility, 62
Bioethics and Safety Act, 188, 192
Bioethics Core Curriculum, 130, 171
Bioethics education, 126, 131, 146, 147, 157, 166, 167, 180, 181, 190
Bioethics infrastructure, 20, 32, 188–189
Bioethics network, 166, 180
Bioethics program, 1, 9, 26, 67, 130, 142, 146, 156, 158, 159, 181, 193
Bioethics-related law, 187
Biomedical research, 2, 20, 32, 35, 38, 41, 47, 61, 87, 132, 138, 140, 141, 155, 156, 167, 177, 178, 181, 186–189
Biomedical research ethic, 20
Biopiracy, 78
Bioterrorism, 5, 9, 32
Biowarfare, 111
Brain drain, 67, 108, 113
Brazil, 7, 105, 163–166, 169
Buddhism, 190

C
Canada, 26, 49, 176
Canterbury *vs.* Spence, 49
Capacity building, 16, 20, 24, 33, 67, 142, 146, 147, 149, 155, 159, 176, 180, 181, 190, 193
Carata Samhita, 185
Care ethics, 4
Caribbean, 164–167, 172
Chester *vs.* Afshar, 49
China, 7, 105, 185, 189, 190, 193

Chiropractic, 75
Climate change, 5, 69, 135
Clinical ethics consultation, 189
Colonization, 164
Commodification, 26, 27
Competent person, 50, 51
Confidentiality, 94, 111, 155, 190
Conflict of interest, 48, 60, 133
Confucianism, 190
Council of Europe, 34, 45, 93, 139–141, 147, 148
Croatia, 142, 147
Cuba, 105, 165, 166
Cultural diversity, 7, 19, 25, 27, 36, 37, 41, 68, 78, 158, 163–172
Cultural pluralism, 30

D
Death with dignity, 192, 193
Declaration of Alma-Ata, 169
Declaration of Helsinki, 9, 34, 35, 138, 169, 177, 188
Declaration of Istanbul, 92, 96
Degradation, 33, 68, 103
Degrowth Declaration, 69
Denmark, 129
Developing countries, 5, 7, 9, 18, 24, 27, 30, 33, 63, 66, 67, 70, 74, 79, 103, 104, 107, 110, 113, 117, 118, 168, 169, 176, 177, 179
Disability, 33, 35, 110
Disaster risk reduction, 69
Discrimination, 31, 34, 39, 73, 77, 78, 82, 83, 86, 88, 91–99, 102, 103, 107, 111, 114, 118, 147, 148, 153, 164, 167–171, 191
Distributive justice, 190
DNA mutation, 109
Doctrine of consent, 47, 48

E
East Asia, 185–193
Egalitarianism, 154
Egypt, 105, 151, 156, 180
Electronic health record, 87
Embryonic stem cell, 48, 168, 191
Encyclopedia of Bioethics, 57
Environmental, 3, 9, 33, 35, 58, 60, 61, 68, 74, 82, 83, 86, 101–104, 106, 108–109, 113, 114, 117, 135, 159, 167, 169–171
Environmental degradation, 33, 68, 103
Equitable societies, 26
Estonia, 84

Ethical challenge, 84, 127, 135, 155
Ethical commitment, 135
Ethical governance, 26, 186
Ethical judgement, 127
Ethical minimum, 6
Ethical oversight, 187, 188
Ethical principle, 3, 6, 19, 64, 76, 94, 146, 147, 157, 160, 169, 187
Ethical value, 146, 148, 154, 155, 158, 170
Ethics education, 119, 126, 131, 147, 157, 159, 166, 167, 178–181, 190
Ethics education program, 24, 146, 159, 180, 190
Ethics Teacher Training program, 142, 146, 159, 180
Euthanasia, 192
Exploitation, 10, 26, 27, 32, 33, 62, 67, 79, 92–95, 98, 102, 107, 108, 153, 177

F
Federico Mayor, 6, 15, 16
Feminist, 51, 190
Food and Agriculture Organization, 30, 143
France, 17, 130, 139, 142, 180
Free radical, 109
Fundamental freedom, 31, 34, 37, 40, 94, 178
Fundamental human right, 61, 64
Future generation, 9, 19, 36, 40, 68, 169

G
Gates Foundation, 65
Gender equality, 69, 170
Genetic information, 88, 192
Genetic literacy, 87
Genetic risk, 86, 87
Genetic screening, 86
Genetic testing, 6, 85, 88, 141
Genome sequencing, 83
Genomic research, 186, 192
Georgia, 145, 146
Germany, 17, 130, 140, 142
Global bioethics, 3, 5–10, 18, 20, 23–24, 57–70, 82, 126, 135, 156, 166, 180
Global challenge, 103
Global Ethics Observatory, 1, 18, 24, 119, 146, 147, 159
Global justice, 10, 32
Global Summit, 131, 142
Globalization, 4, 5, 8–10, 15–16, 18, 19, 60, 66, 156, 170, 175, 181
Grand Challenges, 65, 117, 118

H

Hakim, 154
Health equity, 62, 159
Health risk, 113
Healthcare professional, 27, 47, 171
Hermeneutic ethics, 4
Hinduism, 190
Hippocratic Oath, 154
Human cloning, 17, 186, 188
Human Development Report, 59
Human dignity, 25, 29–31, 34, 36, 37, 40, 44, 45, 47, 66, 67, 94, 97, 130, 148, 149, 153, 157, 188
Human experimentation, 2, 43, 44, 47
Human Genetic Data, 15, 16, 24, 40, 82, 84–87, 147
Human Genome, 2, 6, 7, 15, 16, 24, 34, 40, 63, 67, 84, 86, 87, 141, 147, 153, 155, 166, 192
Human Genome Organisation, 34
Human Genome Project, 2, 87, 166
Human Right, 13, 15–16, 19, 24–27, 29, 31, 34, 39, 40, 44–46, 53, 58, 61, 76, 84–86, 93–95, 98, 118, 125, 131–132, 135, 140, 142, 143, 147, 153, 155, 158, 160, 167, 169–172, 177–179, 189
Human rights law, 6, 31, 37, 40, 41, 47, 48, 94
Human rights violation, 148
Human subject protection, 187
Human trafficking, 93
Human vulnerability, 10, 26, 35, 39, 48, 61, 62, 94, 168, 171, 179
Hunter *vs.* Hanley, 49
Huntington's disease, 85–87

I

Iceland, 84
Immune system, 109
Improper inducement, 32
Incidental finding, 85, 87, 88
India, 7, 105, 161, 185, 189, 191–193
Indigenous people, 74, 78, 79, 164, 165
Indigenous population, 79, 88, 164
Inequality, 8–10, 61, 65, 66, 102, 126, 168
Inequities, 5, 8, 59, 62, 106–108, 110, 112, 164, 168–170
Information technology, 110, 114
Informed consent, 5, 34–38, 45–48, 50, 76, 77, 79, 84, 85, 137, 155, 176, 187, 189, 190

Injustice, 2, 8, 10, 59, 97, 166
Institutional Review Board, 35, 138, 187
Intercultural dialogue, 27
Intergovernmental Bioethics Committee, 1, 18, 24, 25, 141, 158
International Bioethics Committee, 6, 7, 14, 15, 18, 23–24, 30, 35, 36, 44–51, 58, 74, 81, 93, 98, 107, 118, 126, 135, 139, 141, 142, 147, 149, 154, 158, 164, 166, 171, 176, 179, 186
International biomedical research, 188
International community, 17, 33, 41, 63, 114, 193
International human rights law, 6, 31, 36, 37, 40, 41, 46, 48, 94
International moratorium, 113
Iran, 17, 105, 180
Iraq, 151, 152
Islamic Academy of Fiqh, 155
Islamic bioethic, 154–156, 158
Islamic Code of Medical Ethic, 152, 156
Islamic medicine, 152, 156
Islamic Shari'a, 152
Islamic tradition, 190
Italy, 130, 142, 180

J

Japan, 26, 166, 185, 189, 192
Justice, 3, 10, 25, 27, 32, 36, 64, 67, 68, 76–78, 94, 95, 97, 102, 106–108, 127, 148, 153, 155, 157, 164–166, 168, 171, 190

K

Kant, 3, 51
Kazakhstan, 145, 146
Kenya, 177–180
Knowledge Construction, 16, 18
Koïchiro Matsuura, 15, 16, 19
Korea, 185, 186, 188, 189, 191–193
Kuwait, 151, 152, 156

L

Lack of capacity, 39, 52, 106
Latin American bioethic, 164, 168, 172
Lebanon, 151, 156, 159
Libya, 151
Life expectancy, 59, 78
Lithuania, 17, 142
Living Well, 167, 172
Lun ta-i ching-ch'eng', 185

M

Medical ethic, 2, 4, 47, 58, 127, 152, 155, 156
Medical tourism, 5, 9, 170, 188
México, 17, 105, 115, 164–168, 170
Migration, 27, 70, 84, 108, 171
Millennium Development Goal, 68, 103, 117, 170
Minimum prenuptial period, 94
Moral atomism, 51
Moral conscience, 6
Moral imperialism, 7
Moral judgement, 154
Moral pluralism, 172
Moral sensitivity, 148
Morocco, 151, 152
Multiculturalism, 36, 164

N

Nano-based method, 105
Nanobiosensor, 105
Nanodevice, 104, 109, 111
Nanodiplomacy, 116–117
Nano-divide, 107, 112
Nanogovernance, 116–117
Nanopanopticism, 111
Nano-poor region, 108
Nano-rich region, 108
Nanosystem, 103, 104
Nanotechnology, 101–119
Nanotube, 103, 104, 108–110
Nanowaste, 108, 110, 114
Narrative ethic, 4
National Bioethics Committee, 1, 27, 125–135, 148, 155, 157, 159, 176, 178–181, 189, 192
National Ethics Committees, 67, 130, 141, 142, 152, 157
National Ethics Councils, 139, 140
Naturopathy, 75
Neglected disease, 110
Neglected tropical disease, 62, 94
New Zealand, 26
Non-discrimination, 39, 78, 83, 92–95, 98, 103, 147, 168, 171, 191
Nonmaleficence, 4
Non-stigmatization, 39, 83, 88, 92–95, 98, 103, 171, 191
Normative instrument, 1, 6, 7, 13, 14, 17, 127
Nuremberg Code, 34, 43, 44, 137, 154

O

Oman, 151, 159
Organ commercialism, 92, 93, 97
Organ trafficking, 91–95, 97–98, 188, 191
Organ transplantation, 91–99, 155, 156, 186–188, 191
Oriental medicine, 185
Oviedo Convention, 140

P

Palestine, 151, 152
Pan-European Biobanking, 84
Participatory democracy, 172
Patents right, 110
Paternalism, 44
Patient autonomy, 44, 45, 49, 50, 169
Patients' right, 44, 45, 47, 49, 155
Personal integrity, 26, 35, 39, 48, 61, 62, 94, 179
Personalized medicine, 82
Phenomenological ethic, 4
Philippines, 105
Physician-patient relationship, 44
Pluralism, 25, 27, 30, 36, 37, 41, 135, 163–172
Positive obligation, 61
Poverty, 2, 5, 9, 26, 27, 33, 58, 61, 62, 68, 69, 101, 106, 107, 117, 148, 164, 168, 170, 176
Prenatal Diagnostic Techniques, 191
Presumed consent, 96, 97
Principle of autonomy, 34, 44, 47, 76, 77
Principles of bioethics, 27, 34, 44, 76, 149, 153, 155
Principles of justice, 27, 78
Privacy, 31, 34, 92, 111, 155, 159, 188, 192
Private-public partnership, 65
Prophet of Islam, 153
Public discourse, 8, 115
Public education, 99, 147, 168
Public engagement, 115–116, 179

Q

Qatar, 151, 152
Qi gong, 75

R

Reasonable medical practice, 49
REDBIOÉTICA, 166, 167
Reibl v Hughes, 49
Reproductive autonomy, 191
Reproductive cloning, 31, 154
Research ethic, 20, 38, 152, 155
Research Ethics Committees, 126, 129, 138, 156, 166, 169, 176–177, 187–189
Resource allocation, 155, 159, 169, 190

Index

Rhazes, 152
Risks and benefit, 52
Rogers v Whittaker, 49
Romania, 142
Russian, 142, 146

S

Sale of organ, 27
Saudi Arabia, 7, 151, 156, 159
Serbia, 142, 147
Sex selection, 191
Shared commitment, 65
Sharing of benefit, 10, 19, 32, 168
Singapore, 185, 189, 193
Slovakia, 142
Social activism, 186
Social bioethic, 10, 57–70, 171
Social ethic, 168
Social inequalitie, 8, 69
Social inequities, 108, 168
Social justicem, 68
Social responsibility, 9, 19, 32, 33, 39, 47, 48, 58–63, 65, 66, 68, 69, 153, 167, 168, 179
Social vulnerability, 59–62
Solidarity, 1, 8, 10, 27, 32–34, 36, 38, 41, 61, 64–68, 78, 94, 167, 169
South Africa, 7, 105
Special vulnerability, 31, 61, 62, 68
Spiritual therapies, 77
Sri Lanka, 105
Status of Scientific Researcher, 67
Stem Cell scandal, 186
Stigmatization, 82, 86, 91–99, 102, 107, 111, 114, 118
Structural injustice, 8, 10
Surrogacy, 27
Sustainable development, 63, 68, 101–119, 148, 172
Switzerland, 127, 129, 133
Syria, 151

T

Tajikistan, 146
Tanzania, 176, 177
10/90 gap, 176
Thailand, 105, 185, 188
Tai chi, 75
Toxic hazard, 114
Traditional medicine, 45, 73–80, 193
Transcultural context, 36

Transparency, 116, 131, 133, 170
Transplant tourism, 93, 95–98, 191
Tunisia, 151
Turkey, 17

U

Uganda, 177, 180
Ukraine, 146, 147
United Arab Emirates, 151, 156
United Kingdom, 49, 84, 129, 140
United Nations, 8, 14, 19, 23–27, 29, 31, 40, 44, 65, 68, 79, 93, 108, 115, 117, 118, 141, 142, 148, 153, 165, 176
United States, 2, 26, 114, 128, 164, 165
Universal Declaration, 7, 13–20, 24–27, 29–41, 45–46, 50, 53, 58, 61, 63, 76, 86, 93–95, 118, 125–135, 147, 153, 158, 159, 165, 167, 169–172, 177–179, 189, 192
Universal Declaration on the Human Genome and Human Rights, 7, 15, 24, 40, 63, 147, 153, 192
Universalism, 15–16, 63
Utilitarianism, 154

V

Van Rensselaer Potter, 2
Vulnerability, 9, 10, 25–27, 35, 39, 48, 59–62, 68, 94, 102, 107, 168, 169, 171, 179
Vulnerable people, 168
Vulnerable population, 97, 107, 112, 114, 119, 147, 155, 169, 172

W

World Bank, 10, 65
World Health Organization, 24, 25, 30, 58, 69, 74–76, 92, 131, 142, 143, 148, 157, 158, 165, 169, 186, 191
World Medical Association, 44, 138, 148
World Trade Organization, 30
World War II, 43, 47

Y

Yemen, 151, 156
Yoga, 75

Z

Zimbabwe, 177, 180

MIX
Papier aus verantwortungsvollen Quellen
Paper from responsible sources
FSC® C105338

If you have any concerns about our products,
you can contact us on
ProductSafety@springernature.com

In case Publisher is established outside the EU,
the EU authorized representative is:
**Springer Nature Customer Service Center GmbH
Europaplatz 3, 69115 Heidelberg, Germany**

Printed by Libri Plureos GmbH
in Hamburg, Germany